D1064536

MOLECULAR VIROLOGY

MOLECULAR VIROLOGY

C. *Arthur Knight*

Professor of Molecular Biology and
Research Biochemist in the Virus Laboratory
University of California, Berkeley

McGRAW-HILL BOOK COMPANY

New York St. Louis San Francisco Düsseldorf Johannesburg
Kuala Lumpur London Mexico Montreal New Delhi
Panama Paris São Paulo Singapore Sydney Tokyo Toronto

This book was set in Palatino by Monotype Composition Company, Inc.
The editors were William J. Willey and Andrea Stryker-Rodda;
the designer was J. E. O'Connor;
the production supervisor was Bill Greenwood.
The printer was Halliday Lithograph Corporation;
the binder, The Maple Press Company.

MOLECULAR VIROLOGY

2 3 4 5 6 7 8 9 0 HDMM 7 9 8 7 6 5

Library of Congress Cataloging in Publication Data

Knight, Claude Arthur, date
Molecular virology.

1. Virology. 2. Molecular biology. I. Title.
[DNLM: 1. Molecular biology. 2. Virology. QW160 K69m 1974]
QR360.K58 576'.64 73-18331
ISBN 0-07-035112-0
ISBN 0-07-035113-9 (pbk.)

Dedicated to the memory of *Wendell Meredith Stanley,*
1904–1971,
whose early work marked
the beginning of
molecular virology

CONTENTS

PREFACE

Virology began with the concept of viruses as bizarre agents of disease, but has now expanded to the point where the significance of viruses as genetic entities has assumed an importance equal to if not greater than their role in infectious diseases. These two aspects of virology are of course related, for it seems that it is by virtue of genetic interactions between viruses and host cells that viruses cause diseases. These diseases are so widespread that they are found in virtually all forms of life and are characterized by a broad range of symptoms.

The kind of virology which has proved useful in virtually every other aspect of the discipline is that which emphasizes the physical and chemical characteristics of the viruses. For a while after Stanley's crystallization of tobacco mosaic virus, major attention was logically focused on methods for isolating and purifying viruses and on the study of their chemical and physical properties. After a couple of decades, a climax was reached in this type of investigation with the establishment of the genetic primacy of the nucleic acid components of cells and viruses. From this point, interest has steadily grown in the complex interactions of viruses with cellular constituents and in probing these interactions with ever sharper techniques. Hence, the emphasis is presently centered on cell-virus interreactions at the molecular level and may be termed *molecular virology*.

The methods and findings of molecular virology are widely scattered in a literature which is so voluminous that even the experts have difficulty in keeping abreast of developments, to say nothing about critically evaluating them. There seems a need, especially at the undergraduate level, for a simplified version of the principles and implications of virology. Moreover, in contrast to earlier times when it seemed logical to treat viruses in the context of the major classes to which they belonged, i.e., bacterial, animal, or plant, it now seems essential to attempt to integrate knowledge of viruses. This approach makes it feasible to focus attention on the fundamental properties that make a virus a virus without becoming hopelessly engulfed in the vast store of details and esoteric jargon that surround different groups of viruses.

Thus, the primary purpose of this book is to provide for interested undergraduates (or even advanced high school students) a condensed and integrated presentation of basic principles of virology together with guidance to literature which will challenge the most advanced student. In addition, it is hoped that graduate students in a variety of disciplines will find this concise presentation helpful in grasping a subject in which they may have a minimal background or that they will find that it broadens and supplements virology courses already taught with narrower orientations.

The author is greatly indebted to R. C. Williams and H. W. Fisher for many of the electron micrographs and to other virologists as noted for the rest. He also gratefully acknowledges the suggestions and work of E. N. Story in preparing illustrations and the helpful criticisms of the text made by Rosemary Paterson.

C. Arthur Knight

MOLECULAR VIROLOGY

CHAPTER 1 HISTORICAL BACKGROUND

1-1 *Virus Diseases Are Ancient*

Virus diseases have apparently afflicted the peoples of the earth, other animals, and plants for many centuries. How ancient these maladies are is difficult to say, owing to the lack of recorded, reliable diagnoses. Various accounts suggest that smallpox existed in India and China at least 2 or 3 centuries before Christ, but the first authentic account in literature seems to be that of the Persian physician known as Rhazes (A.D. 860–932), who wrote graphic descriptions of the symptoms of both smallpox and measles. In addition to written records, archeologic findings, such as the Egyptian bas-relief shown in Fig. 1-1, point to the early existence of certain human viruses. The chief figure in this stele appears to bear the marks of recovery from crippling poliomyelitis.

Among the viral diseases of domestic animals, rabies is perhaps the oldest for which there are records. This virus disease is commonly associated with rabid dogs, and Aristotle wrote in the fourth century B.C. of the madness and fury of dogs with this disease and their ability

FIG. 1-1. **A bas-relief dating from the eighteenth Egyptian dynasty (ca. 1500 B.C.). Note the shriveled leg of the priest, a characteristic feature of recovery from paralytic poliomyelitis. (From the Ny Carlsberg Glyptothek, Copenhagen.)**

to transmit the affliction to other animals by biting. All warm-blooded animals, including man, are highly susceptible to rabies. In man the disease is often called *hydrophobia* (from the Greek, "fear of water"), and the description given by Celius Aurelianus in the second century A.D. is unmistakable. A portion of his account follows:

> In the beginning of the attack, anxiety with no reason for it, irritability and malaise, restless movement, light sleep and disturbed, insomnia, stretching and continual gaping, and an unremitting desire to vomit, an unusual susceptibility to air, no matter how quietly the patient may have been resting, intoleration and loathing of liquids, little desire to drink. When the disease is established there is thirst and at the same time dread of water, at first at the sight of it, later at the very sound or mention of it. . . .

Records concerning plant, bacterial, and insect virus diseases are much more modern than those which describe human and animal viral afflictions, although this fact by no means indicates that such viruses are more recent in origin. The oldest known plant virus disease appears to be the mosaic disease of tulips, commonly termed "tulip breaking" because of the striking variegations it causes in the flowers (Fig. 1-2). Carolus Clusius (Charles de l'Écluse) described typical variegations of this tulip break disease in 1576, not long after the flower had been introduced to Western Europe from Turkey. No one suspected at that time, nor was it shown until about 3½ centuries later, that the dramatic striping and mottling of tulips was caused by a virus. The more novel the variegation of the flower, the more eagerly it was sought. Nowadays, variegated tulips (except for genetic variants) are usually rogued out of nurseries and destroyed before the mosaic can spread to other stock, particularly to the solid-color tulips. This procedure is in great contrast to the frenzy with which such tulips were sought in the seventeenth century. High sums were paid at the time for certain bulbs or seedlings, and fantastic deals were made for the more bizarre specimens. In Holland in 1625, the following goods were exchanged for just one bulb of the "Viceroy" tulip:

4 tons wheat	4 barrels beer
8 tons rye	2 barrels butter
4 fat oxen	1,000 lb cheese
8 fat pigs	1 bed with accessories
12 fat sheep	1 full-dress suit
2 hogsheads wine	1 silver goblet

FIG. 1-2. **Tulip mosaic. The streaking and feathering of the flowers shown here illustrate some types of "color breaking" which occur in this virus infection, and may be contrasted with the appearance of blossoms from uninfected plants. Also shown are portions of infected and uninfected leaves, the former showing a characteristic mottling pattern. (From McKay and Warner, Natl. Hortic. Mag. 12, 179–216, 1933.)**

The height of this passion, reached in Holland in 1634 to 1637, is described as "tulipomania." The tulipomania period was one of wild speculation in tulip bulbs and even seedlings. Many people expected to get rich by trading in tulips, just as millions in the United States almost 3 centuries later thought to make a fortune in the stock market. Intervention by the government finally stopped the mad tulip gambling, but it is largely to the tulip craze that we are indebted for the copious written records and art that make it possible to link modern knowledge with the old, and to say with assurance that a plant virus disease, tulip mosaic, existed in the seventeenth century.

Virus diseases of insects have probably existed as long as the virus diseases of higher animals and plants. According to Steinhaus, the fact that insects could suffer from disease was known before the Christian era. However, it was not until the middle of the nineteenth century that Cornalia and Maestri described symptoms which can unmistakably be linked to a currently known virus infection, the jaundice or polyhedrosis disease of silkworms.

Since the existence of bacteria was discernible only after the development of the microscope, it is natural that a diseased condition of these microbes was late to be recognized. Presumably, bacteriologists of the nineteenth century observed signs of bacterial virus action in their cultures, but serious consideration of bacterial virus disease began only with the stimulating papers of the British bacteriologist F. W. Twort (1915) and the Canadian Felix d'Herelle (1917). By this time, viruses were already acknowledged as a distinct class of microorganisms.

In summary, despite the absence of corroborating records, it appears from what we know about viruses today that these disease agents have probably existed as long as living forms have. In fact, as will be discussed in a later chapter, there is some speculation that a virus-like object may even have constituted the most primitive form in the evolution of living things.

1-2 Germ Theory of Disease and the Discovery of Viruses as Unique Disease Agents

VIRUSES ARE THE SMALLEST MICROBES

Though infectious diseases have been recognized for centuries, the nature of the causative agents has been revealed only by twentieth-century investigations, and the picture may not yet be complete.

The essential features of our modern germ theory of disease were advanced as early as the seventeenth century by three European scientists, Kircher, Redi, and Leeuwenhoek. Kircher conceived of infectious organisms, *contagio animata*, although he thought they were spontaneously produced in decomposing matter; Redi produced evidence that at least the larger organisms, such as maggots, were not spontaneously produced in decaying matter but originated outside from other living creatures; and Leeuwenhoek, with his ingenious "microscopes," actually saw and described "little animals" (protozoa and bacteria) from the human mouth and the intestine, from seawater, and from many other sources.

Leeuwenhoek's pioneering observations of protozoa and bacteria were made with astonishing accuracy, considering the instruments he employed. His "microscopes," numbering more than 200, were really magnifying glasses consisting of a single biconvex lens fixed between plates of metal. The specimen was fixed to the point of a needle which could be adjusted by a screw arrangement to move the object with respect to the lens. If the object was solid it was glued to the needle point; if liquid, a small drop was placed on a piece of thin, blown glass which was glued to the needle tip. The object was viewed by looking through the lens toward a light. A picture of a replica of one of Leeuwenhoek's magnifying glasses (the best ones magnified objects about 200 times) is shown in Fig. 1-3. Among the multitude

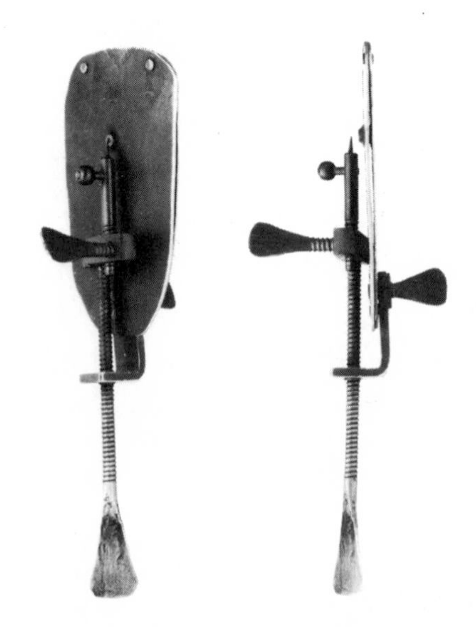

FIG. 1-3. **A Leeuwenhoek "microscope," showing front and side views. (The microscope pictured is a replica manufactured in a shop at Delft located on the site of Leeuwenhoek's dwelling.)**

of objects examined by Leeuwenhoek was some of the white material from between his teeth. In a letter written to the secretary of the Royal Society in London on September 17, 1683, he described what he had seen "with great wonder, that in the said matter there were many very little living animalcules, very prettily a-moving." From his descriptions and drawings of these particular animalcules it seems clear that Leeuwenhoek was describing common oral bacteria, including a motile bacillus, *Selenomonas sputigena,* micrococci, *Leptothrix buccalis,* and a spirochete, probably *Spirochaeta buccalis.*

However, the significance of Leeuwenhoek's observations was not grasped at the time, and for 2 centuries progress in understanding infectious disease was retarded by a preoccupation with the epidemic potentialities of the atmosphere (miasma and effluvia). Efforts were concentrated on means to disinfect the air in times of epidemics. Aromatic fumigants and vinegar, fumes of mineral acids, the lighting of fires, and firing of cannon were all considered effective against contagion at one time or another. This state of affairs is illustrated by the action taken in Philadelphia against yellow fever during a frightful epidemic in 1794. About 10 percent of the population was wiped out by this dread disease, and during the height of the panic which spread through the city the College of Physicians issued a manifesto of suggestions which stated the opinion that fires (which were lighted everywhere in the streets in the supposition that they would purify the air) were an ineffectual, if not a dangerous, means of checking the progress of this fever. It was recommended instead that more dependence be placed upon the burning of gunpowder. This suggestion was widely accepted, but the boom of the cannon and the crack of muskets only punctuated the rattle of the carts carrying off the dead. Furthermore, the noise and danger were so great that the mayor shortly after forbade this practice.

In the nineteenth century, renewed consideration was given to the idea of parasitic cause of disease. Jacob Henle in Göttingen, Germany, conceived of infectious agents with properties very much like those presently attributed to viruses, but his idea of a microbial cause of disease failed to gain general acceptance until the subsequent work of three brilliant investigators. One of these, Robert Koch (1843–1910), was a student of Henle's. The other two were Louis Pasteur (1822–1895) and Joseph Lister (1827–1912).

Koch made his first major contribution to understanding parasitic disease as a country doctor in a German village. In a remarkable series of studies, he worked out the life history of the anthrax bacillus, including a demonstration of the significance of the heat-resistant spore stage of this organism. From this accomplishment, Koch went on to demonstrate the nature and the disease-producing capacity of other pathogenic bacteria, such as the cholera vibrio and the tubercle bacillus. His work on tuberculosis is considered particularly outstanding. Koch is also noted for his development of basic techniques in bacteriology—solid-culture and hanging-drop-culture methods, staining with aniline dyes, and photomicrography.

Nine great discoveries of Pasteur are commemorated on the arches of his tomb in the crypt of the Pasteur Institute in Paris, and all but three of these relate to the development of the germ theory of disease. Not only did Pasteur help to disprove decisively the classical notion of spontaneous generation, but he also showed by actual isolation and description that specific bacteria were the causes of several infectious diseases and of the spoilage of beer and wine. Pasteur also worked with virus diseases, notably silkworm jaundice and rabies, and demonstrated their microbial, infectious nature, but did not distinguish them from bacterial infections.

Lister was a surgeon in England who was dismayed at the high incidence of fatal infections following surgery. For example, almost half the amputees failed to survive, owing to subsequent infections. In a flash of inspiration, Lister reasoned that the cause of gangrenous wounds might be, like Pasteur's spoilage of wine, attributable to a microbe. If so, it would be desirable to sterilize surgical wounds. The application of heat was not as practical as it was in Pasteur's work; however, Lister had learned about the successful use of carbolic acid (phenol) in disinfecting sewage, and decided to try this for sterilizing wounds. The experiment was highly successful, and antiseptic surgery was born.

Thus the work of Pasteur, Koch, Lister, and their followers brought in the Golden Age of Bacteriology. Among the diseases soon identified as caused by bacteria were leprosy, gonorrhea, typhoid fever, lobar pneumonia, glanders, erysipelas, diphtheria, tetanus, plague, and bacillary dysentery. It seemed as though the causes of all infectious diseases had been found.

However, near the end of the nineteenth century, some experiments were performed with diseased tobacco plants which were destined to cause a significant revision in the germ theory of disease.

In 1876, Adolf Mayer, a German agricultural chemist, was called from the University of Heidelberg to be director of the Agricultural Experiment Station at Wageningen, Holland. While in Holland his attention was called to a condition of tobacco which was variously called "bunt," "rust," or "smut." Though all these names had some relation to symptoms seen at one time or another in the tobacco fields, a more generally characteristic symptom was the pattern of light and dark green areas seen on infected leaves, which led Mayer to propose in 1886 the name "mosaic disease of tobacco." Mayer demonstrated by analysis of leaves and soil that the disease could not be properly attributed to a mineral imbalance. This led him to do a key experiment. He ground up a diseased leaf with a little water, sucked some of the resulting green emulsion into fine capillary tubes, and used these tubes to prick the leaf veins of healthy plants. In 9 cases out of 10, the inoculated healthy plants became heavily diseased. This established the infectious quality of mosaic disease and led to a search for the causative microorganism using the newly developed techniques of Koch and others. This search failed; no bacterium or fungus could be causally associated with the disease. Nevertheless, Mayer concluded that the disease might be a bacterial one, whose specific nature would be revealed by future studies.

Such studies were made a few years later by a young Russian scientist, Dmitrii Ivanovsky, investigating tobacco diseases in the Crimea. At the end of a short report on mosaic disease presented to the St. Petersburg Academy of Science on February 12, 1892, Ivanovsky made the striking remark, "I have found that the sap of leaves attacked by the mosaic disease retains its infectious qualities even after filtration through Chamberland filter-candles." This observation suggested a disease agent smaller than anything known before, but Ivanovsky apparently was not confident enough of his findings to go on to such a revolutionary conclusion, for immediately following the above, he said:

> According to the opinions prevalent today, it seems to me that the latter is to be explained most simply by the assumption of a toxin secreted by the bacteria present, which is dissolved in the filtered sap. Besides this there is another equally acceptable explanation possible, namely, that the bacteria of the tobacco plant penetrated through the pores of the Chamberland filter-candles, even though before every experiment I checked the filter used in the usual manner and convinced myself of the absence of fine leaks and openings. [James Johnson translation.]

Six years later (1898), the Netherlands scientist Martinus W. Beijerinck repeated the Ivanovsky experiment by passing sap expressed from the leaves of mosaic-diseased tobacco through a filter candle of the sort shown in Fig. 1-4, and showing that the filtrate was still infectious. Beijerinck was confident that his filter was holding

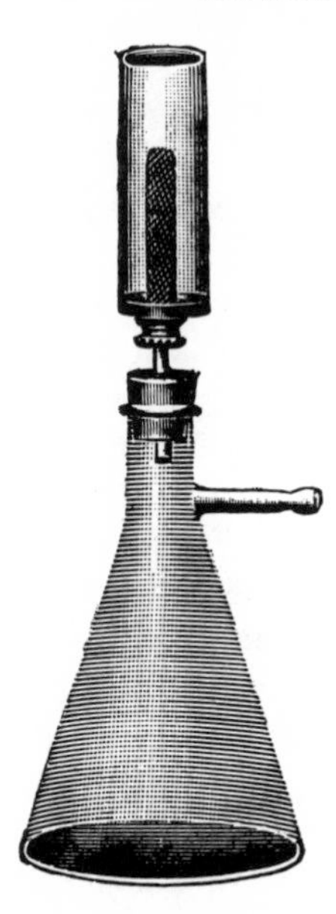

FIG. 1-4. **A filtering apparatus of the type employed by Ivanovsky and Beijerinck.**

back bacteria. Furthermore, the infectious agent was shown to be smaller than common bacteria by placing sap on the surface of blocks of agar gel and finding that the agent diffused through the gel at a measurable rate rather than remaining on the surface as bacteria do. To this unprecedentedly small pathogen Beijerinck applied the terms *contagium vivum fluidum* ("contagious living fluid") or "virus." Beijerinck went on to demonstrate that that virus retained its infectivity for more than 2 years in dried diseased leaves or in the form of infectious sap dried on pieces of filter paper. He noted also that infectivity could be precipitated from sap of diseased plants by treatment with alcohol and that this precipitate, dried at 40°C, maintained infectivity. Infectivity of sap was shown to be abolished by boiling or by treatment with formaldehyde. Beijerinck concluded his historic paper with the prediction that many other "nonparasitic" (meaning nonbacterial or nonfungal) diseases of unknown cause, such as peach yellows and peach rosette, would be found to be caused by viruses.

Ivanovsky was still not convinced that tobacco mosaic was caused by the newly recognized class of disease agents proclaimed by Beijerinck, and in 1899 published a paper describing the results of additional experiments on mosaic disease, in which he concluded, "Admittedly the experiments are not very numerous and the percentage of diseased plants small; nevertheless I believe that the bacterial nature of the infection is scarcely to be doubted."

MARTINUS WILLEM BEIJERINCK

DMITRII IVANOVSKY

ADOLF MAYER

FIG. 1-5. **Three pioneer tobacco mosaic virologists: Mayer, Ivanovsky, and Beijerinck. (From Phytopathological Classics, No. 7, Cayuga Press, Inc., Ithaca, N.Y., 1942.)**

In the meantime, Loeffler and Frosch (1898), working with foot-and-mouth disease, found that lymph containing the infectious agent lost no infectivity upon passage through filters which presumably retained the smallest bacteria. This suggested a small size for the agent of foot-and-mouth disease, and the possibility was considered that it might be a bacterial toxin present in the lymph. However, after serial passage through as many as six animals, the lymph of the last animal was still infectious, a result which seemed explicable only on the basis of a replicating infectious agent. In discussing their findings, Loeffler and Frosch noted that there were other diseases of unknown cause, such as smallpox, cowpox, scarlet fever, measles, and rinderpest, and that perhaps these, too, were caused by exceedingly small organisms like the foot-and-mouth disease agent.

In any case, at the turn of the twentieth century and for the next 30 years, many infectious agents were tested for their filterability. As a consequence, agents causing such diverse diseases as yellow fever, rabies, Rous sarcoma of chickens, rabbit myxomatosis, infectious lysis of bacteria, cucumber mosaic, potato X disease, and many others were classified in the newly recognized group of "filterable viruses." The latter term became the name for this class of microorganism, but with time it was recognized that some viruses, though small enough, did not pass through bacterial filters readily, owing to charge and adsorption effects. Also other characteristics, such as sedimentation behavior, came to be used more and more to demonstrate the small size of viruses. Finally, with the development of the electron microscope (ca. 1940), it became possible to identify virus particles with objects of definite sizes and shapes. Thus the term *filterable* was eventually dropped, although the characteristic small size (now definable in precise terms), which filterability had indicated, remains a distinguishing quality of viruses, along with infectious capacity and ability to reproduce only within living cells.

1-3 Dawn of the Chemical Era of Virology

CHEMISTRY AND PHYSICS WERE SLOW TO ASSERT THEMSELVES

The horizons of virology expanded considerably in the first three decades of the twentieth century as disease after disease was demonstrated to be of viral origin. Maladies as diverse as Rous sarcoma of chickens and infectious lysis of bacteria were shown to fall in the category of viral infections. In general, it was necessary only to show that bacterial-free filtrates of macerated tissue were infectious to determine that a disease was of viral etiology.

Along with the rapid expansion of the list of virus diseases, numerous studies were made of the specific pathologic effects caused by viruses and of the ways in which viruses are transmitted. In order better to characterize these newly recognized disease agents, many studies were also made of the effect of various chemical and physical agents on infectivity. An excellent review of these pioneer biochemical investigations on viruses is available in the article by Stanley in *Handbuch der Virusforschung*.

Some major conclusions which could be drawn from the early biochemical work were that protein denaturants, formaldehyde, oxidizing agents, strong acids or bases, and high temperatures are inimical to viruses, whereas the milder protein precipitants, low temperatures, and neutral pH generally are not. These results, along with others, suggested that virus activity might be associated with proteinaceous material.

In 1928 and 1929, Purdy, at the Boyce Thompson Institute, reported the production in rabbits of antisera to extracts of healthy tobacco and of mosaic-diseased tobacco (i.e., infected with tobacco mosaic virus). Common antigens (substances eliciting the formation of antibodies and reacting in serologic tests) were demonstrable in the two types of extracts, but it was shown by complement fixation and precipitin tests made with adsorbed sera that there was some antigenic material in the extracts from mosaic-diseased plants which was not present in healthy tobacco juice. It was further shown that this specific substance was present in extracts of tomato, pepper, and petunia if the plants had been infected with tobacco mosaic virus (TMV), but was not present in tobacco plants infected with other viruses. Significantly, the antiserum to TMV extract displayed a strong neutralizing power when mixed with infectious extracts, whereas antiserum to healthy tobacco caused only a slight reduction in infectivity of viral extracts. Furthermore, it was repeatedly shown that a quantitative relation exists between the antigenic content of virus extracts and the infectivity as measured by local lesion counts obtained on the appropriate type of tobacco. These and rather extensive immunologic studies made on TMV extracts by Matsumoto and colleagues at Taihoku Imperial University in 1930 to 1933 led Beale (née Purdy) to conclude that the specific antigenic material in TMV extract "may be the virus itself." Since proteins were known to be the major antigenic substances of nature, the implication regarding TMV is obvious.

In the period 1927 to 1931 Vinson and Petre at the Boyce Thompson Institute came close to obtaining purified TMV in a series of experiments in which they precipitated the infectious material from extracts of mosaic-diseased plants by treatment with safranin, acetone, or ethyl alcohol. As the infectious fraction was separated from the bulk of impurities associated with it, an increase in its relative nitrogen content was observed. This, considered with the precipitability of the infectivity by protein precipitants and the observation that the infectious principle moved in an electric field after the manner of proteins, indicated that the virus might be a protein. This series of investigations was climaxed by the reported production of infectious, crystalline preparations, although these were not obtained consistently, were only "moderately active" (infectious), and contained 33 percent ash (largely calcium oxide). Thus Vinson and Petre succeeded in showing that TMV behaves as a chemical substance, but their attempts to demonstrate conclusively the chemical nature of the virus fell short of success, owing to persistent impurities in their preparations, uncertain biologic assays, and variable large losses of virus in the purification procedures.

In England, in about 1935, F. C. Bawden began an impressive series of studies on plant viruses with a report on "the relationship between the serological reactions and the infectivity of potato virus X." Using infective material which had been precipitated by treatment of leaf extracts with carbon dioxide and resuspended in dilute salt, he found, as Beale had shown for the TMV system, that a close relationship exists between the antigen content of X-virus preparations (as measured by their optimal flocculation points with antisera) and their virus content (as measured by the local lesion method). Furthermore, it was shown that destruction of infectivity by heat, aging, and alcohol was accompanied by the loss of flocculating power with antiviral sera. However, inactivation by phenol greatly reduced the serologic flocculation but did not completely destroy it, while inactivation by formalin left the flocculating power unimpaired. These results, like those of Beale, were generally suggestive that viral activity is closely associated with protein.

An important clue to the shape of TMV was obtained in about 1932 by Takahashi and Rawlins at the University of California. Duggar and Karrer had found some 11 years earlier by filtration methods that the virus has an apparent diameter of approximately 30 nm. Filtration measurements, however, could tell almost nothing about shape. Therefore, little progress was made in this direction until Takahashi and Rawlins became interested in the use of stream double refraction (also called *flow birefringence* or *anisotropy of flow*) for detecting asymmetric particles of colloidal dimensions. Rod-shaped particles show strong flow birefringence (i.e., the rods, transiently oriented in a flowing stream and viewed through crossed Nicol prisms or crossed sheets of Polaroid, transmit flashes of light). Disk-shaped particles are much weaker in this respect, and spheroidal particles, being completely symmetric, do not show the phenomenon. Juice from healthy plants was found to exhibit no double refraction when forced from a small tube, but juice extracted from various parts of infected tobacco or tomato plants showed the same double refraction as obtained with colloidal vanadium pentoxide and other suspensions of rodlike particles. From these experiments, the conclusion was drawn: "The evidence therefore indicates that the virus of tobacco mosaic, or some substance regularly associated with it, is probably composed of rod-shaped particles." Since these tests were made on crude extracts containing the virus, the possibility could not be excluded that the doubly refracting material was a product of infection distinct from the virus itself. Hence the observation failed to receive the attention it deserved until some years later, when it was shown to be valid for the highly purified virus as well. Tobacco mosaic virus does, of course, occur as rodlike particles in nature.

At about this same period (1934) some important observations were made by Max Schlesinger, working on bacteriophages at the Institut für Kolloidforschung in Frankfurt, Germany. Schlesinger found that a purified phage preparation which by serologic tests was free from host cell material nevertheless gave strong color tests for protein and contained phosphorus. This led him to suggest that nucleoprotein might be a major constituent of bacterial viruses. The proposal lacked the force which it might have carried had the presence of purine and pyrimidine bases been demonstrated as well as phosphorus, and, in any case, the significance of nucleic acids was not at this time adequately appreciated. Moreover, there was also some question whether or not bacteriophages were chemically and physically in the same class with plant and animal viruses. Hence the proposal that viruses might be nucleoproteins was not generally accepted until several years later.

CRYSTALLIZATION OF TOBACCO MOSAIC VIRUS. STANLEY EXPLODES A GERM-THEORY MYTH

In 1933, Wendell M. Stanley began studies on the chemical nature of TMV. He first repeated some of the work of Vinson and Petre, gradually modifying the fractionating methods, with especial attention to control of pH at the various steps. Infectivity was determined on various fractions with unprecedented accuracy by use of the local-lesion assay method newly developed by Francis Holmes at the Boyce Thompson Institute in New York State and studied in detail by the Australian workers Samuel and Bald. Stanley also had available some of the crystalline proteolytic enzymes prepared by J. H. Northrop and associates, and had a personal familiarity with the methods involved in their preparation. In one crucial experiment of a series on the effect of chemical reagents on viral activity, Stanley found that the infectivity of TMV was largely destroyed by pepsin at a pH at which the virus was otherwise stable. This led him to state, "It seems difficult to avoid the conclusion that tobacco mosaic virus is a protein, or closely associated with a protein, which may be hydrolyzed with pepsin."

Subsequently, Stanley combined repeated precipitation with ammonium sulfate with decolorization by treatment with lead subacetate to obtain high yields of purified virus. Such virus in aqueous solution was caused to crystallize in the paracrystalline state by adding sufficient saturated ammonium sulfate to cause turbidity, followed by the gradual addition, with stirring, of 0.5 saturated ammonium sulfate in 5 percent acetic acid. The needle-like paracrystals obtained were like those shown in Fig. 1-6. These were infectious at dilutions as high as 10^{-9} gm per ml, and the infectivity of the material, in contrast to what Vinson and Petre had found, was not lost by as many as 10 successive recrystallizations. From the results of many kinds of tests, the crystalline material appeared to be protein, and preliminary osmotic pressure and diffusion measurements indicated that this protein had an extraordinary molecular weight, of the order of several millions. The infectivity of the preparations was shown to depend on the integrity of the protein, and hence infectivity could be considered a property of the protein. Stanley concluded his historic paper, published in *Science* in 1935, with the statement, "Tobacco-mosaic virus is regarded as an autocatalytic protein which, for the present, may be assumed to require the presence of living cells for multiplication."

Subsequently, numerous other viruses were obtained in crystalline form, although it was about 20 years after the crystallization of TMV before the first animal virus (poliovirus) was crystallized. Some virus crystals are illustrated in Fig. 1-6. Electron micrographs of some virus crystals are especially interesting because they demonstrate directly the beautifully regular, three-dimensional array of particles in a crystal. Such a micrograph is shown in Fig. 1-7.

In the original work, Stanley did not detect phosphorus in his virus preparations and considered the active material to be a crystal-

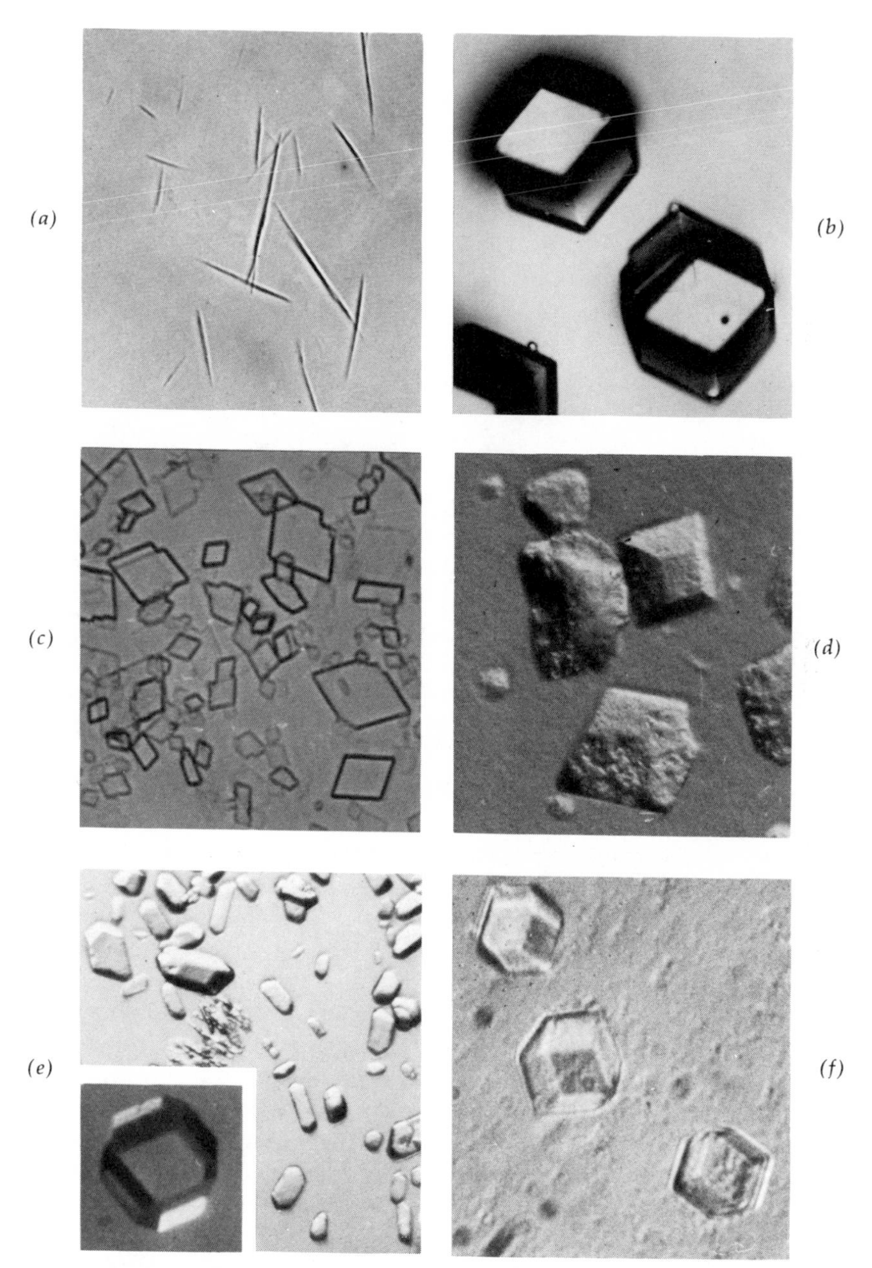

FIG. 1-6. **Some virus crystals: (a) tobacco mosaic virus; (b) tomato bushy stunt virus; (c) southern bean mosaic virus; (d) turnip yellow mosaic virus; (e) poliovirus (courtesy of F. L. Schaffer); (f) mouse polyoma virus (courtesy of W. T. Murakami).**

FIG. 1-7. **An electron micrograph of a crystal of tobacco necrosis virus. (Courtesy of Ralph W. G. Wyckoff.)**

lizable protein. However, Bawden, Pirie, and associates made the important discovery in 1936 that purified strains of TMV contain phosphorus and carbohydrate, and that these components are present in the form of ribonucleic acid, which can be released from the virus by heat denaturation. This finding was confirmed by Stanley shortly afterward, and although he at first viewed the nucleic acid as probably not essential for infectivity, he later reversed his judgment, and together with others established that several different plant viruses could be isolated as nucleoproteins.

The discovery of the chemical nature of TMV had a tremendous impact on medical and biologic science. It was revolutionary to think of a crystallizable, infectious disease agent, but once the idea was accepted it led to the investigation of the chemical and physical properties of a host of other viruses. Moreover, a number of scientists were stimulated to begin detailed investigations of the infectious process on the physical and chemical level, and others began explorations of the chemical changes accompanying mutation of a virus, thus helping to create a new field, chemical genetics. The availability of purified virus preparations even played a role in the development of the electron microscope, for viruses proved to be objects of ideal size to use in testing and perfecting the operation of this newly contrived instrument. The development of other physical tools, such as the ultracentrifuge, was also greatly influenced by application to the study of purified viruses. Thus the general significance of TMV has been much greater than its importance as a plant pathogen.

1-4 Development of Molecular Virology

FINDINGS FROM STUDIES ON ANIMAL, BACTERIAL, AND PLANT VIRUSES HAVE COMBINED TO YIELD SIGNIFICANT ADVANCES BOTH IN VIROLOGY AND IN MOLECULAR BIOLOGY

Subsequent to the early biochemical studies of plant viruses by Stanley and others, several key developments occurred which accelerated the growth of molecular virology.

Around 1940 the electron microscope was used in Germany by G. A. Kausche and G. Melchers and colleagues and in America by Stanley and T. F. Anderson to visualize viruses for the first time. Ever since, this tool has been an invaluable aid in determining the structure (see Chap. 3) and deducing the function of all kinds of viruses.

At about this same time a new area of great importance in virology developed, namely, the study of the process of infection employing *Escherichia coli* bacterial cells and certain viruses (called *bacteriophages* or *phages*) which preyed upon them. Studies which purposed to exploit the interaction of single virus particles with single bacterial cells were enthusiastically performed by members of a clique of scientists known as the Phage Group. This group was led by such men as M. Delbrück, A. D. Hershey, S. E. Luria, and M. Demerec. Starting with the experimental definition of the one-step growth cycle and elucidation of distinctive stages of this cycle, both the group and its significant findings grew in size. The investigations of the Phage Group, though varied, had a definite genetic bias, which, however, was balanced by strongly biochemically oriented work led by S. S. Cohen, E. Evans, F. Putnam, L. Kozloff, T. T. Puck, and their associates. Over the next two or three decades, studies of the phage-bacterium interaction provided models for investigation of the process of infection with other types of viruses and at the same time, like those of the chemistry of plant viruses, provided knowledge of general importance to molecular biology as well. These advances, some of which will be considered in later chapters, include: the discovery that DNA (deoxyribonucleic acid) is the genetic material of phages containing this type of nucleic acid; that viral nucleic acid can be integrated into host cell DNA; that cellular genes can be transported by a virus from one cell to another; that infection can cause the induction of new, virus-specific enzymes in cells; the demonstration of messenger RNA (ribonucleic acid); the understanding of some mechanisms for the regulation of transcription and translation of genetic messages; and many aspects of molecular genetics.

Inspired by the success of the phage workers, researchers in animal virology turned to the development of techniques for growing animal cells in culture in order to analyze the process of infection free from some of the complicating factors of the whole organism. Even before this, notable advances in animal virology were made by employing a simple host, the chick embryo, which proved to be susceptible to a wide variety of viruses and to be convenient to work with. The potentiality of the chick embryo was first demonstrated by E. W. Goodpasture and his colleagues, and its use was greatly forwarded by W. I. B. Beveridge and F. M. Burnet. Influenza viruses were among the many viruses cultured successfully in the chick embryo, and studies on the infectious process with these viruses were considerably advanced by the discovery of an accurate quantitative assay. This was the finding of G. K. Hirst, since extended to many other animal viruses, that influenza viruses agglutinate red cells in proportion to concentration of virus, and that the degree of agglutination is accurately measurable in a simple densitometer. Another quantitative assay that grew out of culture of viruses in the chick embryo was based on the development of characteristic spots (foci or pocks) on the chorioallantoic membrane, the number and character of these spots depending on the virus and its concentration (see Beveridge and Burnet). However, the demonstration by R. Dulbecco of plaques caused by animal viruses on monolayer cell cultures superseded the pock assay and proved to be a stimulus to the systematic analysis of infection with animal viruses. Some basic findings which have issued from the study of animal viruses in cell cultures include the recognition of novel paths for transcription (e.g., that involving a "reverse transcriptase" enzyme), and translation (e.g., the synthesis of large polypeptide precursors which are later cleaved to several proteins with specific functions); the distinctive host ranges shown by whole virus compared with viral nucleic acid; the transformation of cells with respect to growth habits and morphologic characteristics caused by certain viruses; the fusion of cells caused by infection with some viruses, with resultant combination of nuclei; the maturation of enveloped viruses at cell membranes accompanied by incorporation of host constituents into the virus particles; the segmentation of some viral nucleic acids and genotypic mixing; and polarity of viral nucleic acids with respect to need, or lack of it, to be transcribed before translation can occur.

Through all the exciting developments with bacterial and animal viruses there continued to run an unbroken strand of fundamental discoveries with plant viruses. These include specific modifications of plant viruses which provided a basis for the rational development of inactive animal virus vaccines; the first direct demonstration of infectious nucleic acid; the first reconstitution of infectious virus from component parts; the demonstration of viral protein subunits, a forerunner of quaternary (subunit) structure of proteins in general; the first illustration of the translation of virus mutation to the structure of a gene product, the coat protein of TMV, whose primary structure was also the first of any virus to be completely determined; extension of the genetic code to the plant kingdom by showing that changes in viral coat protein coincided with predictions based on the assumed code and alterations in the nucleic acid induced by a specific mutagenic chemical; the existence of satellite viruses and viruses with multiparticulate genomes; and viroids, very small infectious RNA molecules.

Thus molecular virology flourished in the middle of the twentieth century. As noted by W. K. Joklik, advances in our understanding of each of the major classes of viruses have been profoundly dependent on discoveries concerning the others. This catalytic effect is still evident.

References

BOOKS

Beveridge, W. I. B., and F. M. Burnet: *The Cultivation of Viruses and Rickettsiae in the Chick Embryo,* Med. Res. Council Special Report Series No. 256, London, 1946.

Burnet, F. M.: *Principles of Animal Virology*, 2d ed., Academic, New York, 1960.

Cairns, J., G. S. Stent, and J. D. Watson: *Phage and the Origins of Molecular Biology*, Cold Spring Harbor Laboratory of Quantitative Biology, Cold Spring, N.Y., 1966.

Clendening, L.: *Source Book of Medical History*, Dover, New York, 1942.

Cohen, S. S.: *Virus-induced Enzymes*, Columbia, New York, 1968.

Dobell, C.: *Antony van Leeuwenhoek and His "Little Animals": A Collection of Writings by the Father of Protozoology and Bacteriology*, Dover, New York, 1932.

Major, R. H.: *Classic Descriptions of Disease*, 3d ed., Charles C Thomas, Springfield, Ill., 1945.

Phytopathological Classic No. 7: *Mayer, 1886; Ivanowski, 1892; Beijerinck, 1898; Baur, 1904, Three Early Papers on Tobacco Mosaic and One on Infectious Variegation*, translated from the German by James Johnson, Cayuga Press, Inc., Ithaca, N.Y., 1942.

Steinhaus, E. A.: *Principles of Insect Pathology*, McGraw-Hill, New York, 1949.

Van Iterson, G., Jr., L. E. Den Dooren De Jong, and A. J. Kluyver: *Martinus Willem Beijerinck, His Life and His Work*, Martinus Nijhoff, The Hague, 1940.

JOURNAL ARTICLES AND REVIEW PAPERS

Bawden, F. C.: The Relationship between the Serological Reactions and the Infectivity of Potato Virus "X," *Br. J. Exp. Pathol.* **16**:435–443, 1935.

———, N. W. Pirie, J. D. Bernal, and I. Fankuchen: Liquid Crystalline Substances from Virus-infected Plants, *Nature* (London) **138**:1051, 1936.

Beale, H. P.: The Serum Reactions as an Aid in the Study of Filterable Viruses of Plants, *Contrib. Boyce Thompson Inst.* **6**:407–435, 1934.

Dulbecco, R.: Production of Plaques in Monolayer Tissue Cultures by Single Particles of an Animal Virus, *Proc. Natl. Acad. Sci., U.S.A.* **38**:747–752, 1952.

Frylink, A.: The Tulip. Part I. Its Early History, *Garden J., N.Y. Bot. Garden* **4**:5–15, 1954.

d'Herelle, F.: Sur un microbe invisible antagoniste des bacilles dysentériques, *C. R.* **165**:373–375, 1917.

Hirst, G. K.: The Quantitative Determination of Influenza Virus and Antibodies by Means of Red Cell Agglutination, *J. Exp. Med.* **75**:49–64, 1942.

Kausche, G. A., and H. Ruska: Die Struktur der "kristallinen Aggregate" des Tabakomsaikvirusproteins, *Biochem. Z.* **303**:221–230, 1939.

McKay, M. B., and M. F. Warner: Historical Sketch of Tulip Mosaic or Breaking, the Oldest Known Plant Virus Disease, *Nat. Hortic. Mag.* **12**: 179–216, 1933.

Melchers, G., G. Schramm, H. Trurnit, and H. Friedrich-Freksa: Die biologische, chemische und elektronenmikroskopische Untersuchung eines Mosaikvirus aus Tomaten, *Biol. Zentralbl.* **60**:524–556, 1940.

Purdy, H. A.: Immunologic Reactions with Tobacco Mosaic Virus, *J. Exp. Med.* **49**:919–935, 1929.

Schlesinger, M.: Zur Frage der chemischen Zusammensetzung des Bakteriophagen, *Biochem. Z.* **273**:306–311, 1934.

Stanley, W. M.: Isolation of a Crystalline Protein Possessing Properties of Tobacco Mosaic Virus, *Science* **81**:644–645, 1935.

———: Biochemistry and Biophysics of Viruses. I. Inactivation of Viruses by Different Agents, in *Handbuch der Virusforschung,* 1st half, R. Doerr and C. Hallauer (eds.), pp. 447–458, Springer-Verlag, Vienna, 1938.

——— and T. F. Anderson: A Study of Purified Viruses with the Electron Microscope, *J. Biol. Chem.* **139**:325–338, 1941.

Takahashi, W. N., and T. E. Rawlins: Method for Determining Shape of Colloidal Particles; Application in Study of Tobacco Mosaic Virus, *Proc. Soc. Exp. Biol. Med.* **30**:155–157, 1932.

Twort, F. W.: An Investigation on the Nature of Ultramicroscope Viruses, *Lancet* **2**:1241–1243, 1915.

Vinson, C. G., and A. W. Petre: Mosaic Disease of Tobacco, *Bot. Gaz.* **87**: 14–38, 1929.

CHAPTER 2 PURIFICATION OF VIRUSES

Purification of viruses was and is fundamental to the determination of essential chemical and morphologic features of viruses. More importantly, qualities of viruses that are significant in their interactions with cellular constituents are brought out by reflection on the principles and techniques that have been applied in the purification of viruses. Moreover, the methods briefly sketched in this chapter (references given at the end of the chapter may be consulted for additional details) are also used in experiments designed to elucidate the infectious process with viruses. Therefore, some grasp of the techniques should enhance appreciation of these experiments. On the practical side, purification of viruses is important because the only feasible defense against animal viruses at present is vaccination, and this procedure has hazards which are reduced by use of appropriately purified viruses.

Although hundreds of virus diseases are recognized, relatively few of the viruses have been isolated in a highly purified state. The obvious reason for this situation is that viruses multiply only within living cells, and the disruption of such cells to release the virus also releases a multitude of cellular constituents, some of which may closely resemble the virus in size and other properties. Moreover, virus particles may adsorb, or be adsorbed on, cellular components, either before or during the isolation procedure. To make matters worse, virus particles may aggregate among themselves to form clumps that are easily lost in cellular debris during the initial steps of purification. Evidence suggests also that some viruses may exist only in relatively unstable forms, perhaps as free nucleic acid.

In any case, the problems of purification are intensified in inverse proportion to the virus yield. In many instances of low virus productivity, the task becomes the proverbial one of looking for a needle in the haystack, with the complication that the needle often has an affinity for the hay. Despite these unfavorable circumstances, some viruses have been obtained in a highly purified state.

Viruses have individual characteristics, and the ease or difficulty with which a given virus can be isolated and purified is related to its properties as well as to the nature of the host and the culture conditions. Even strains of the same virus may differ substantially in ease of purification. Consequently, it is not possible to outline a purification procedure that will work with equal effectiveness for all viruses. However, since the basic problems summarized above are essentially the same for all viruses, certain approaches found to be successful for some viruses are in principle more or less applicable to the purification of viruses in general.

Two common properties of viruses are the basis for purification methods: (1) the exterior surfaces of most viruses consist mainly and sometimes entirely of protein; and (2) viruses are of such size, shape, and density that they are sedimented at about 40,000 times gravity (40,000 g). The high protein content of viruses means that the general techniques of protein fractionation can often be used to advantage in the purification of viruses, while the sedimentation characteristics of viruses frequently enable their separation from salts, proteins, and a wide variety of cell constituents.

In order to determine the effectiveness of any purification procedure, it is essential that an accurate quantitative test for virus infectivity be available. For example, if a virus assay is subject to 50 percent or more variability, not uncommon in biologic testing, it is usually not possible to determine with certainty the fraction in which the virus is contained or the extent to which the purification procedures are destroying virus activity. Thus it was of crucial importance in Stanley's purification and crystallization of tobacco mosaic virus to make use of the Holmes local-lesion assay method, rather than the older, imprecise assay by dilution end point.

When a sufficiently accurate measure of infectivity is available, it is possible to adjust purification conditions to stay within the bounds of the pH and thermal stabilities of the virus and to evaluate the effects of salts and other chemicals. If information concerning these conditions is lacking, it is customary to begin by performing all operations in the cold (4°C) and at a pH of about 7. Potassium phosphate buffer, 0.01 to 0.1 *M*, has proved to be a good ionic medium for several viruses. Salt mixtures, such as Ringer's solution, are needlessly complex media for most viruses; on the other hand, unbuffered "physiologic" saline solution is deleterious to some viruses, owing to its tendency to be somewhat acidic.

Among the various techniques for isolating and purifying viruses, centrifugation is most often applied. In some cases, a virus can be concentrated and purified from tissue extracts by means of centrifugation alone. In many cases, however, centrifugation needs to be supplemented by other methods. Only a brief sketch of useful techniques will be given here, but references to some excellent detailed reviews of these methods will be included at the end of the chapter.

2-1 *Differential Centrifugation and Density-Gradient Centrifugation*

THE MOST REFINED OF THESE TECHNIQUES ARE EXQUISITELY SELECTIVE

The sizes and densities of most known viruses are such that they can be sedimented from solution in an hour or two in centrifugal fields of 40,000 to 100,000 g. Such centrifugal fields are readily obtained in commercially available centrifuges whose rotors are driven by electric motors.

Some outstanding features of motor-driven centrifuges are their compactness, simplicity of maintenance, reliability, and safety. The centrifuge rotors are made of an aluminum alloy so that they are strong but relatively light (see Fig. 2-1). A single machine accommodates a variety of rotors, most of which hold 10 or more plastic tubes. The tubes are held with their bottoms farther from the rotor axis than their tops, thus enhancing the pelleting of material by allowing it to accumulate on the outer tube wall and slide to the bottom. An "angle centrifuge" has its rotors so constructed. Tubes are available in a variety of sizes to match the rotors; as little as 2 ml, or as much as

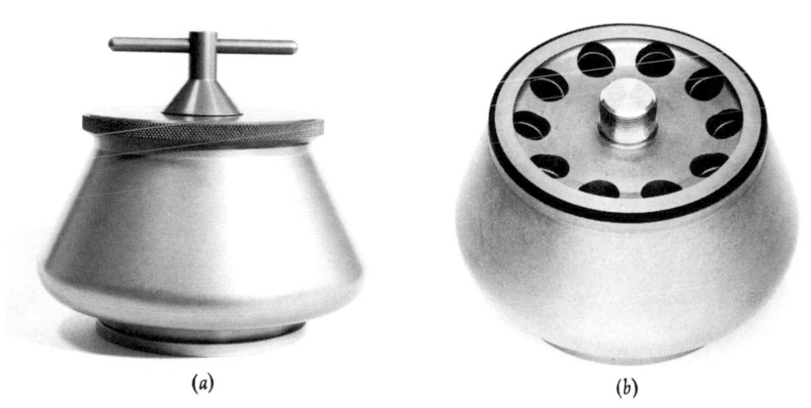

(a) (b)

FIG. 2-1. **A rotor used in high-speed centrifugation. The rotor is machined from die-forged aluminum alloy with controlled grain structure for maximum strength and long life. (a) the rotor with lid in place and ready to be set on the drive shaft of a high-speed motor. Rotor lids are provided with gaskets for maintaining a tight seal and holding contents at atmospheric pressure while the rotor is operated in a vacuum. (b) rotor with the lid removed, showing the cavities in which plastic, sealed tubes are placed for centrifugation. (Courtesy of the Spinco Division of Beckman Instruments, Inc.)**

940 ml, can be handled in one run. Both tubes and rotors are sealed, and since the rotor spins in a vacuum, the initial temperature of rotor and sample changes very little during a run. A refrigeration unit around the walls of the vacuum chamber makes it possible to avoid even slight rises in temperature. The force fields obtainable in one commercial model (see Fig. 2-2) range from 59,000 g in the 21,000-rpm rotor, holding 940 ml, to 368,400 g in the 65,000-rpm rotor, holding 108 ml. A cardinal feature of the electrically driven centrifuge is the flexible shaft supporting the rotor. Such a flexible shaft makes it unnecessary to balance the tubes carefully before centrifugation, inasmuch as the spinning rotor seeks its own axis of rotation.

Preparations of some viruses can be obtained in a highly purified state by use only of the machine just described. The process of "differential centrifugation" means simply the application of alternate cycles of low-speed and high-speed centrifugation. Such alternation can be done in the same centrifuge, although a separate and simpler angle centrifuge is commonly used for the low-speed cycles. In this process, the virus and other materials are sedimented at high speed, but, when the process is operating favorably, only cellular debris

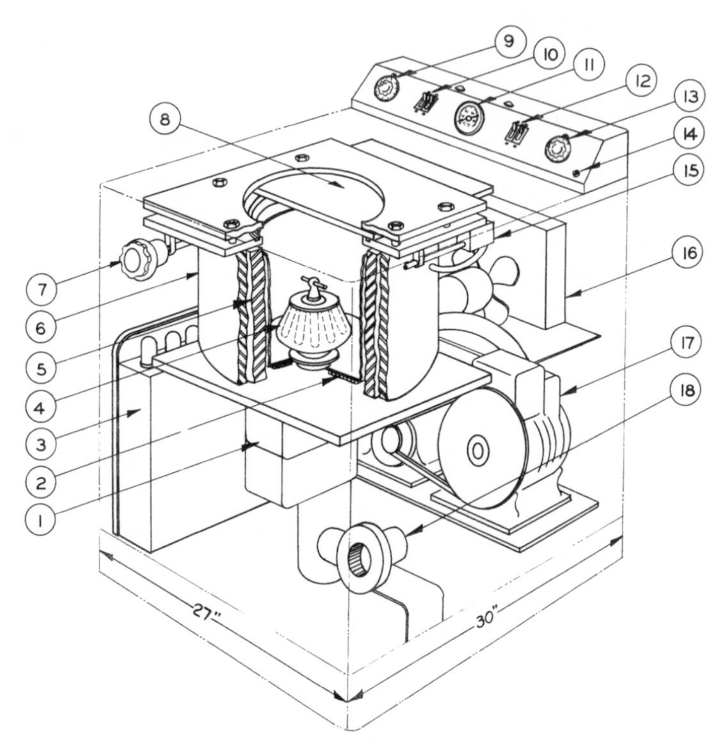

FIG. 2-2. **Schematic perspective of Spinco model L ultracentrifuge. (Courtesy of Spinco Division of Beckman Instruments, Inc.)**

1. Drive unit
2. Refrigeration liner
3. Electronic chassis
4. Rotor
5. Armor plate guard ring
6. Rotor chamber wall
7. Left control knob
8. Rotor chamber door
9. Time clock
10. Main power switch (left)
    Main refrigerator
    switch (right)
11. Tachometer
12. Vacuum pump switch
    (left)
    Brake switch (right)
13. Speed selector
14. Start button
15. Right control knob shaft
16. Refrigerator
17. Vacuum pump
18. Blower

and such materials are removed in the low-speed centrifugations. Hence the virus is purified by employing several cycles of low-speed and high-speed centrifugation.

A technique for preparative microcentrifugation of viruses and other entities of similar size has been described by R. C. Backus and R. C. Williams. In this method, pellets obtained by use of conventional centrifugation equipment are resuspended in 0.01 to 0.1 ml of diluent and are then transferred and sealed into "field-aligning" glass or quartz capsules. These are suspended in a solvent of suitable density in a standard plastic centrifuge tube and are centrifuged in an angle rotor at an appropriate speed (Fig. 2-3). With supplementary equipment, such as a spectrophotometer, these capsules can also be used for analytic ultracentrifugation of virus preparations.

The technique of density-gradient centrifugation, introduced into virology by M. K. Brakke in 1951, has become a powerful adjunct to conventional differential centrifugation for the purification of viruses and related materials. Separations can be made by this method which are either impossible or less sharp in ordinary sedimentation. Furthermore, it is possible under appropriate conditions to use the method to estimate densities and sedimentation coefficients.

The principle of density-gradient centrifugation is the separation of particles partly or entirely on the basis of their densities in a

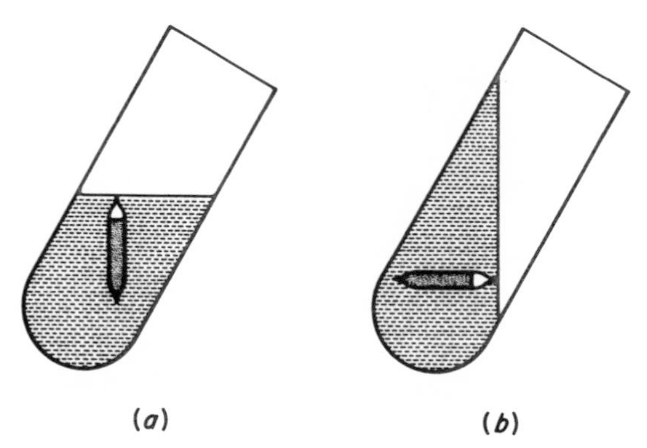

FIG. 2-3. **Microcentrifugation in field-aligning capsules. Capsule floating in centrifuge tube partially filled with salt solution: (a) rotor at rest; (b) rotor at speed. For high-speed centrifugation the tube must be filled by floating mineral oil over the salt solution. (Adapted from Backus and Williams, Science 117, 221–223, 1953.)**

convection-free medium. Such separation can be accomplished in a variety of ways, which differ mainly in such details as the nature of the material used to form the gradient, use of a preformed gradient or one formed during sedimentation, the magnitude of the gravitational field, and whether or not centrifugation is continued until equilibrium conditions are reached.

For convenience in classification, the two main modifications of this type of centrifugation are (1) rate-zonal (or velocity) density-gradient centrifugation, and (2) equilibrium (or isopycnic) density-gradient centrifugation.

In *rate-zonal centrifugation* the virus solution is layered on top of a preformed gradient of sucrose or glycerol in a plastic tube, and centrifuged in a swinging-bucket rotor (Fig. 2-4) for 0.5 to 3 hr at about 90,000 to 420,000 g. This type of rotor allows the axis of the centrifuge tubes to become horizontal at high centrifugal fields. At the conclusion of the run, the zones of different classes of particles can be visualized by use of proper illumination, owing to light scattering ("Tyndall effect") (Fig. 2-5).

Rate-zonal centrifugation has also been termed *gradient differential centrifugation,* because the time of centrifugation and the gravitational field are approximately equivalent to those employed in the

FIG. 2-4. **A swinging-bucket rotor for use in density-gradient centrifugation. Three buckets accommodate cellulose nitrate, polypropylene, stainless steel, or quartz tubes. One bucket and its shaft are shown removed. (Courtesy of Spinco Division of Beckman Instruments, Inc.)**

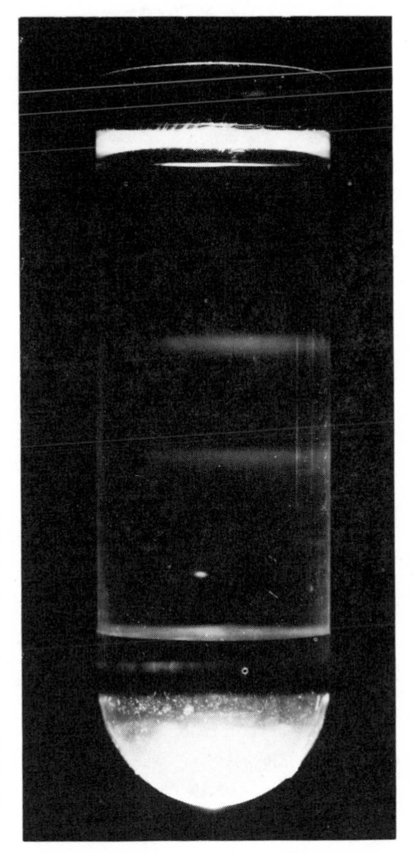

FIG. 2-5. **Rate-zonal centrifugation of a mixture of tomato bushy stunt and tobacco mosaic viruses, each at 0.1 mg per ml. A gradient of glycerol ranging from 10 to 40 percent glycerol 0.1 M phosphate at pH 7 was used. The top band contains bushy stunt virus, and the bottom band, TMV. Centrifugation was for 2 hr in a Spinco SW 25.1 swinging-bucket rotor at a maximum centrifugal force of 90,000 g.**

high-speed cycle of ordinary differential centrifugation. However, a sharp distinction can be made between gradient differential and ordinary differential centrifugation with respect to the location and movement of the sedimenting particles at various stages of the sedimentation process.

In ordinary differential centrifugation in an angle centrifuge, the particles are uniformly dispersed throughout the centrifuge tube at the beginning; during the run they sediment independently to the sides of the tube and thence slide down to the bottom, where they form a gelatinous pellet. Since all classes of particles are found near the bottom of the tube as well as elsewhere, sedimentation for any period of time long enough to produce a pellet will result in a pellet consisting of a mixture of all the particles present. The various particles will appear in the pellet in quantities proportional to their sedimentation rates. The sedimentation rates depend on size, shape, and density of the particles, and the viscosity of the medium. At the conclusion of the run there is nothing to prevent incompletely sedimented particles as well as those in the surface of the pellet from being redistributed by convective disturbances.

In gradient differential centrifugation the mixture is floated as a layer on top of a density-gradient column, and the different classes of particles are sedimented through the column in zones in accordance with the sedimentation rate of each class of particle (hence the term *rate-zonal*). The zones are stabilized against convective disturbances during the run, and at the conclusion, by the presence of the gradient.

The high resolution of rate-zonal centrifugation as compared with differential centrifugation may be illustrated by the separation of brome mosaic and tobacco mosaic viruses. After one rate-zonal centrifugation in a sucrose gradient the two viruses appear in separate bands. Samples from the brome mosaic virus zone have no detectable TMV in them, and samples from the TMV band have about 0.1 percent the original concentration of brome mosaic. Brome mosaic virus sediments in dilute salt about 41 percent as fast as does TMV, and it can be estimated that at least seven cycles of differential centrifugation would be required to obtain the degree of separation accomplished in just one cycle of rate-zonal centrifugation.

If rate-zonal centrifugation is extended for a period of hours, most of the particles reach and remain in zones corresponding to their densities (isopycnic position). The process then is more properly called *equilibrium-zonal centrifugation,* since the zones obtained are essentially the equilibrium ones with respect to the densities of the particles. However, equilibrium-zonal or isopycnic gradient centrifugation is generally set up in a somewhat different manner. In *isopycnic gradient centrifugation,* the density gradient of the suspending medium is formed either prior to or during centrifugation, and the virus solution may or may not be layered at the top of the density-gradient column. Inorganic salts, such as cesium chloride, rubidium chloride, or potassium bromide, are used, usually at concentrations of 6 to 9 *M,* and a smooth gradient develops as a consequence of partial sedimentation of the salt in the centrifugal field. The virus solution may be introduced either before or after formation of the gradient. Centrifugation is then continued for 12 to 24 hr until the particles reach their isopycnic positions, where their densities coincide with those of the suspending medium.

Use of the isopycnic gradient centrifugation method was greatly stimulated by the report of M. Meselson and colleagues in 1957 that they were able by this means to separate the normal DNA of T2 bacteriophage from T2 DNA in which some of the thymine had been replaced by the denser 5-bromouracil. The effective densities of these two types of DNA were 1.7 and about 1.8, respectively. However, the method has been used to detect density differences between particles of less than 0.001 gm per cm^3.

The isopycnic type of density-gradient centrifugation has been employed in the purification of various animal, bacterial, and plant viruses and has also been used to distinguish between strains of some viruses, or, in many cases, to separate the complete and incomplete forms of viruses.

The various forms of density-gradient centrifugation are among the most useful techniques for the purification of viruses and related components. Particularly effective is the successive use of rate-zonal and equilibrium conditions. Contaminants differing from the virus in sedimentation velocity are removed in the first step, while impurities differing from the virus in density are eliminated in the equilibrium centrifugation. Early concern about possible deleterious effects of the gradient-forming materials on viruses or nucleic acids has mainly proved unwarranted. Most viruses and nucleic acids appear quite stable at low temperature in sucrose, glycerol, and strong cesium chloride and similar salts.

The usefulness of the density-gradient principle has been considerably expanded by development of modifications such as density-gradient runs in fixed-angle rotors or in continuous-flow rotors (see Chap. 5 in Maramorosch and Koprowski, *Methods in Virology*, vol. II).

2-2 *Precipitation Methods*

THESE METHODS ARE QUICK BUT ROUGH

Ammonium sulfate at various concentrations is the salt most commonly used in protein fractionation. Like salts in general, ammonium sulfate affects the charge and hydration of proteins, often diminishing their solubility. The degree to which solubility is diminished varies with the nature of the protein, and this variation provides the basis for fractionation of different proteins, or viruses. In some cases, magnesium sulfate is substituted for the ammonium salt, and occasionally ethanol is used, either to precipitate the virus or, by selection of proper conditions, to precipitate impurities from the virus solution. One procedure that has been used in the purification of turnip yellow mosaic virus involves flocculation of plant proteins from the expressed, infectious plant sap by adding ethanol to a concentration of

20 percent, followed by precipitation of virus from the supernatant fluid by addition of a half volume of saturated ammonium sulfate.

The concentration of some animal viruses, such as influenza, simian virus 5, and Newcastle disease viruses, is readily effected by addition of an equal volume of saturated, neutralized ammonium sulfate to infectious chick embryo allantoic fluid or to tissue culture media containing the virus. The precipitate of crude virus thus obtained is usually purified by resuspension in a small amount of buffer or buffered medium and subjection to density-gradient centrifugation. The virus separates from other materials and appears in the gradient in a visible band.

Basic proteins, such as protamine, have been used in the purification of some animal viruses. In certain cases the protamine causes mainly the virus to precipitate, whereas in other instances the protamine flocculates contaminating proteins, leaving the virus largely in the supernatant fluid.

The concentration of large quantities of influenza virus from chick embryo allantoic fluids and their partial purification can be achieved by precipitation at $-5°C$ with 25 to 35 percent methanol.

In general, precipitation methods are used as adjuncts to other procedures, primarily as a means of concentrating viruses from dilute suspensions. The main advantages of the precipitation method are simplicity and rapidity; its chief disadvantage, aside from the question of efficiency (which constitutes a separate problem for each virus), is the possibility that the precipitating agent may cause denaturation of the viral protein. Such denaturation is usually manifested by a failure of the precipitated virus to resuspend properly in water or dilute salt solutions.

2-3 Adsorption Methods

THESE METHODS ARE HIGHLY DIVERSIFIED BUT SELDOM HIGHLY SPECIFIC

Adsorption techniques were among the earliest methods used for the concentration and purification of animal viruses. For example, the virus of foot-and-mouth disease was purified by L. W. Janssen in 1937 by repeated adsorption on, and elution from, finely suspended calcium sulfate. However, for many years adsorption methods were only slightly used. Later, stimulated by the development of new adsorbents, such as ion exchange resins, and cellulose derivatives such as diethylaminoethylcellulose and carboxymethylcellulose, this approach has acquired fresh potentialities. Adsorbents which have been used over the years in the purification of animal and plant viruses include aluminum phosphate, calcium phosphate, ferric and aluminum hydroxides, kaolin, diatomaceous earth, activated charcoal, erythrocytes, alumina, and chitin. In short, almost any available adsorbent may be useful for the purification of viruses.

Sometimes the impurities, rather than the virus, are adsorbed. For example, activated charcoal has been used with considerable success in removing the colored matter from crude preparations of plant viruses. The advantage of adsorption methods is that in individual cases they may offer a selectivity not afforded by other means. The main disadvantages are that, in general, adsorption may not be selective enough to aid substantially in the purification process, and that the rather high salt concentrations sometimes required for elution may be deleterious to the virus. Since materials such as agar gels or dextrans are now known to remove salts rapidly from virus solutions, the latter disadvantage is becoming less important. In practice, adsorption is seldom used as the sole process for purification of a virus. It is almost always coupled with other techniques, most commonly with differential or density gradient centrifugation.

The adsorption technique can be operated as a batch process, which means that virus extract and adsorbent are mixed in a beaker or other container and, after adsorption has taken place, the adsorbent is separated from the rest of the extract by filtration or centrifugation. Subsequently, virus is eluted from adsorbent by treatment with an appropriate salt solution; if impurities, rather than virus, have been adsorbed from the extract, the adsorbent is discarded and the virus in the supernatant is concentrated by centrifugation or some other means. Many workers find it convenient to use columns packed with adsorbent rather than the batch process. This is usually called *column chromatography*, the term chromatography having originated from the early application of the method to the separation of colored materials (chromophores) on such columns.

2-4 *Treatment with Enzymes*

MANY VIRUSES ARE RESISTANT TO PROTEASES AND NUCLEASES, SO THAT NONVIRAL PROTEINS AND NUCLEIC ACIDS CAN BE DIGESTED AWAY FROM VIRUS PARTICLES

Treatment of crude virus preparations with proteolytic enzymes and nucleases is often useful in purification procedures. Two noteworthy properties possessed by most viruses make such treatments feasible:

(1) Viral proteins are not as readily denatured as are common cellular proteins and hence resist digestion by proteolytic enzymes under conditions which permit digestion of contaminant proteins. For example, snails can feed on mosaic-diseased tobacco leaves and excrete undegraded TMV. This situation is indicative of a general feature of animal and human viral diseases, namely, that viruses ingested in the food can escape destruction in the alimentary tract. In fact, some viruses can institute infections in the cells of the digestive tract. Moreover, active virus, with the capacity to initiate new infections, often remains in the excreta. Such conditions constitute important reservoirs of infection in nature. (2) The nucleic acids of virus particles are either surrounded by protein or deeply embedded in it, so that free nucleic acid in or around virus particles may be degraded with nucleases without damaging the viral nucleic acids. After treatment with enzymes, the macromolecular viral particles are readily separated from the low molecular weight digestion products of proteins and nucleic acids by differential centrifugation or other procedures.

2-5 Extraction with Organic Solvents

IN SOME CASES ORGANIC SOLVENTS DISSOLVE AWAY FATTY OR COLORED IMPURITIES OR DENATURE NONVIRAL PROTEINS, THUS AIDING PURIFICATION OF A VIRUS

Another purification procedure which depends considerably on the resistance to denaturation of viruses as compared with common cellular proteins is extraction with organic solvents. An example of this is afforded by one process of purification of poliovirus, in which a helpful step is an extraction with *n*-butanol. After an aqueous solution of crude poliovirus is shaken with *n*-butanol, three layers are observed. The top, alcoholic layer contains fatty extractives; the virus remains largely in the lower aqueous solution; and many of the nonviral proteins are denatured and appear in the interface. The same sort of treatment but employing a mixture of equal volumes of *n*-butanol and chloroform is useful in the purification of some plant viruses. In this case, chlorophyll and other pigmented materials appear as a gelatinous scum at the interface, and the virus is found in the aqueous top layer. In place of the organic solvents just mentioned, certain fluorocarbons, such as Freon 112 (FCl_2C-CCl_2F) or Genetron 226 (F_2ClC-CCl_2F) alone or mixed with *n*-heptane, can be used. In some cases at least, fluorocarbons seem to be unusually efficient in gathering and holding nonviral proteins and lipids in the organic phase while keeping the viruses, without significant loss of infectivity, in the aqueous phase.

In general, the efficacy of solvent extraction in the purification of viruses depends considerably on finding the right solvent and conditions, and especially on the degree to which nonviral components are removed or denatured, preferentially to the virus, by the extraction. Obviously those viruses, such as influenza and other myxoviruses, whose particles are held together at least in part by fatty constituents, cannot be treated with lipid solvents without disintegration of structure. In the successful application of extraction procedures the virus is usually recovered from the aqueous phase and further purified by differential centrifugation.

2-6 *Treatment with Antiserum*

OCCASIONALLY A VIRUS IS PURIFIED BY TREATING CRUDE PREPARATIONS WITH SERUM THAT COMBINES WITH AND PRECIPITATES IMPURITIES. IF ANTISERUM TO VIRUS IS AVAILABLE IT CAN BE USED TO PRECIPITATE THE VIRUS, LEAVING IMPURITIES BEHIND

Serologic techniques are used occasionally in the purification of viruses. In some cases, sera directed against host proteins are employed to precipitate the contaminating proteins, after which the virus is freed of serum elements by two or more cycles of differential centrifugation. In other instances, antiserum to the virus is added to form an insoluble virus-antibody complex. This complex, removed from the surrounding medium and washed a few times with dilute salt, can be dissociated by tryptic digestion or by fluorocarbon extraction. The virus can then be recovered by centrifugation or other techniques.

One of the difficulties in using antiserum to remove impurities is to get serum that is potent enough or that contains antibodies in optimal proportions to form precipitates of the various contaminating proteins.

2-7 *Electrophoresis*

PARTICLES DIFFERING IN CHARGE FROM THE VIRUS PARTICLES CAN BE SEPARATED FROM VIRUS BY ELECTROPHORESIS

The movement of charged particles in solution under an impressed electric field is called *electrophoresis*. Electrophoresis is sometimes used in the purification of viruses, but has found limited application because of the complexity of equipment required, difficulties in sampling, or inability of the procedure to handle desired quantities of material. Viruses and cellular proteins usually bear different and distinctive net charges at a given pH, allowing a separation of the two on the basis of their different migration rates in an electric field. Electrophoresis can be performed in salt solutions, density gradients, or on various solid media such as gels or paper strips. Theoretically the method has considerable potential, and further developments may increase its usefulness in virus purification.

2-8 *Liquid Two-Phase Systems*

VIRUSES CAN BE SIMULTANEOUSLY CONCENTRATED AND PURIFIED

P. Albertsson and colleagues developed an unusual fractionating system based on partition of small particles in liquid two-phase systems. Water-soluble pairs of polymers, such as dextran-methylcellulose or dextran-polyethylene glycol, can be used to concentrate and at least partially purify animal, bacterial, and plant viruses. Concentration of a virus by this technique is dependent upon the unequal distribution (low partition coefficient) of the virus between the two phases, and upon adjustment of phase volumes in a way such that most of the virus goes into the small-volume phase. With proper adjustment, concentrations as high as 100-fold can be obtained in one step, or as high as 10,000-fold in a two-step operation. Purification occurs when nonviral materials distribute differently from the virus, as they often do, and can be enhanced by repeated partition, as, for example, in a countercurrent distribution apparatus. The usefulness of the method is obviously extended if it is coupled with other purification procedures such as extraction with organic solvents, centrifugation, etc.

The liquid two-phase separation is a simple, mild treatment by means of which rather great concentrations of virus can be achieved. Virus purification by this means is significant, but as with many other techniques, it requires application of supplemental methods if highly purified virus is the objective. A considerable problem is the separation of virus and the polymer of the partitioning system after phase separation is achieved. Virus and polymer can sometimes be separated by repeated ultracentrifugation, which sediments mainly the virus, but separation is usually more efficiently achieved by passage of the virus-containing phase through a column of agar or dextran gel. The latter process of molecular sieving can be very useful in virology, not only for the application just mentioned but also for removal of salts and impurities that are of lower molecular weight than the virus, and for separation of macromolecules of various sizes, as indicated below.

2-9 Gel Chromatography

SIEVING IS A GENTLE WAY TO SEPARATE PARTICLES OF DIFFERENT SIZES

The use of gel chromatography, also called *gel filtration* and *molecular sieving,* to purify viruses was popularized around 1962 by Steere and associates, using very small particles of agar for the sieving material. Now, refined agar preparations called *agarose,* in the form of fine spherical beads, are used, as well as similar preparations of dextrans and polyacrylamides. Gel beads of whatever kind can be prepared with different average porosities. When a mixture of particles of various sizes is passed through a column of gel, smaller particles penetrate the gel and their passage is retarded, whereas larger particles are excluded and pass on to be eluted from the column ahead of the smaller particles. In this manner a significant fractionation of particles according to size can be achieved. Consequently, the method is useful in separating virus particles from impurities when the latter differ sufficiently in size from the virus; or the procedure can be used, as Steere demonstrated, to separate virus particles themselves into classes according to size (e.g., tobacco mosaic virus particles of different lengths). The method is of even greater usefulness in separating viral proteins from one another and in determining their approximate molecular weights.

2-10 *Purity and Homogeneity*

SOME SEMANTICS AND SOME PRAGMATISM

The operations just described on the preceding pages are performed with the intent of extricating a virus from infected cells. The resulting product is a virus preparation whose quality needs, for some purposes at least, to be defined. Infectivity tests are usually made; a positive result indicates, by definition, that one or more viruses are present, presumably as particles which may be experimentally identified. The essence of identification is the association of infectivity with a class of particles uniquely described by a number of physical parameters, e.g., size, shape, density, particle weight, and electrophoretic mobility. The degree to which a virus is identified depends upon the number of agreements between physical properties and infectivity, upon the accuracy of the determinations of physical characteristics, and upon the sensitivity of the infectivity measurement. The quality of the identified virus is then often evaluated by referring to the "purity" or "homogeneity" of the virus preparation. Both these terms are limited in usefulness in ways that need to be understood.

The difficulty with the adjective "pure" when applied to a virus preparation is that it often implies too much in too general a sense. It might be argued that a virus preparation is pure only if it contains particles that are identical in every conceivable and measurable respect. In this case the demonstration of purity is an almost hopeless task. On the other hand, a pure virus preparation might be thought of less rigorously as one that is demonstrably homogeneous, i.e., consists of similar particles. Here purity and homogeneity become synonymous. But then the questions arise, homogeneous in what respect and to what degree? It soon becomes clear that the adjective "pure" should be abandoned in favor of specific quantitative evaluations of homogeneity or inhomogeneity. Thus it is possible to "purify" a virus and to obtain a "highly purified preparation," but it is most difficult, if possible at all, to demonstrate a "pure" virus preparation.

What about "homogeneity"? This word denotes a degree of uniformity, but only in regard to the criteria used for detecting lack of uniformity. Some means of evaluating the homogeneity of virus preparations include the following: (1) sedimentation in the ultracentrifuge, which permits evaluation of classes of particles in terms of size, shape, and density; (2) examination in the electron microscope, which provides data on the numbers of particles of specific sizes and shapes; (3) measurement of electrophoretic properties, by means of which the uniformity of particles with respect to surface charge can be evaluated; (4) serologic tests, especially those made in such a way as to reveal the presence of host cell constituents if they are present; (5) infectivity measurements, especially when they can be related to the number of particles present; (6) x-ray crystallography, which is applicable under some circumstances to the evaluation of regularity of the internal structure; (7) solubility tests, by means of which the uniformity of the particles with respect to those surface groupings which interact with solvent can be evaluated; (8) crystallizability, which in some cases reflects the uniformity with respect to size, shape, and nativity (lack of denaturation) of the particles present.

An important aspect of the investigation of virus preparations for homogeneity is the reliability of the tests for homogeneity. The questions should always be asked: What precisely can this measurement be expected to show, and especially, what are the limits to which inhomogeneities can be detected by this method? To the degree that there is uncertainty in the answers to these questions, there must also be reservations about the homogeneity of the virus preparation.

A further caution about the use of the word "homogeneous" has to do with the potential ambiguity of the term. A virus preparation is clearly inhomogeneous when it contains particles which are noninfectious and which are also chemically, physically, and serologically distinct from the infectious particles. However, it is entirely possible for a virus preparation to be homogeneous in one or more respects and inhomogeneous in others. For example, a purified preparation of turnip yellow mosaic virus (TYMV) may contain two major types of particles. The main distinction between them is that one class of particle consists only of protein shells which are noninfectious, whereas the other class contains 32 percent RNA within the same kind of protein shell and these particles are infectious. The particles in such a preparation of TYMV are homogeneous, within the limitations of the methods applied, with respect to crystallizability, electrophoretic behavior, size and shape, proportions of amino acids present, and in ability to bind antibodies. However, they are inhomogeneous with respect to density (the nucleic acid–containing particles are denser) and infectivity (only the nucleic acid–containing particles are infectious, although it is not yet known whether or not these particles are uniformly infectious).

Many other examples of the sort given above could be cited, but these should be sufficient to emphasize that the characterization of a virus preparation in terms of homogeneity is significant only if its types of homogeneity are clearly specified. To this information must be added an evaluation of the limits of inhomogeneity detectable by the tests of homogeneity applied.

References

BOOKS

Albertsson, P.: *Partition of Cell Particles and Macromolecules,* 2d ed., Wiley, New York, 1971.

Habel, K., and N. P. Salzman: *Fundamental Techniques in Virology,* Academic, New York, 1969.

Knight, C. A.: *Chemistry of Viruses,* Springer-Verlag, New York, 1974.

Maramorosch, K., and H. Koprowski: *Methods in Virology,* vol. II, Academic, New York, 1967.

JOURNAL ARTICLES AND REVIEW PAPERS

Adsorption and Adsorption Chromatography

Philipson, L.: Chromatography and Membrane Separation, in *Methods in Virology,* K. Maramorosch and H. Koprowski (eds.), vol. II, pp. 179–233, Academic, New York, 1967.

Venekamp, J. H.: Chromatographic Purification of Plant Viruses, in *Principles and Techniques in Plant Virology,* C. I. Kado and H. O. Agrawal (eds.), pp. 369–389, Van Nostrand Reinhold, New York, 1972.

Centrifugation

Anderson, N. G., and G. B. Cline: New Centrifugal Methods for Virus Isolation, in *Methods in Virology,* K. Maramorosch and H. Koprowski (eds.), vol. II, pp. 137–178, Academic, New York, 1967.

Backus, R. C., and R. C. Williams: Centrifugation in Field-aligning Capsules: Preparative Microcentrifugation, *Science* **117**:221–223, 1953.

Brakke, M. K.: Density-gradient Centrifugation, in *Methods in Virology,* K. Maramorosch and H. Koprowski (eds.), vol. II, pp. 119–136, Academic, New York, 1967.

Markham, R.: The Ultracentrifuge, in *Methods in Virology,* K. Maramorosch and H. Koprowski (eds.), vol. II, pp. 3–39, Academic, New York, 1967.

Meselson, M., F. W. Stahl, and J. Vinograd: Equilibrium Sedimentation of Macromolecules in Density Gradients, *Proc. Natl. Acad. Sci. U.S.A.* **43**: 581–588, 1957.

Electrophoresis

Polson, A., and B. Russell: Electrophoresis of Viruses, in *Methods in Virology,* K. Maramorosch and H. Koprowski (eds.), vol. II, pp. 391–426, Academic, New York, 1967.

Van Regenmortel, M. H. V.: Electrophoresis, in *Principles and Techniques in Plant Virology,* C. I. Kado and H. O. Agrawal (eds.), pp. 390–412, Van Nostrand Reinhold, New York, 1972.

Extraction with Organic Solvents

Francki, R. I. B.: Purification of Viruses, in *Principles and Techniques in Plant Virology,* C. I. Kado and H. O. Agrawal (eds.), pp. 309–310, Van Nostrand Reinhold, New York, 1972.

Philipson, L.: Water–Organic Solvent Phase Systems, in *Methods in Virology,* K. Maramorosch and H. Koprowski (eds.), vol. II, pp. 235–244, Academic, New York, 1967.

Gel Chromatography or Molecular Sieving

Ackers, G. K., and R. L. Steere: Molecular Sieve Methods, in *Methods in Virology,* K. Maramorosch and H. Koprowski (eds.), vol. II, pp. 325–365, Academic, New York, 1967.

Precipitation

Schwerdt, C. E.: Chemical and Physical Methods, in *Viral and Rickettsial Infections of Man,* 4th ed., F. L. Horsfall, Jr., and I. Tamm (eds.), p. 20, Lippincott, Philadelphia, 1965.

Steere, R. L.: The Purification of Plant Viruses, *Adv. Virus Res.* **6**:23–26, 1959.

Purity

Pirie, N. W.: The Criteria of Purity Used in the Study of Large Molecules of Biological Origin, *Biol. Rev.* **15**:377–404, 1940.

Treatment with Antisera

Cramer, R.: Purification of Animal Viruses, in *Techniques in Experimental Virology,* R. J. C. Harris (ed.), p. 165, Academic, New York, 1964.

Treatment with Enzymes

Schwerdt, C. E.: op. cit., p. 21.

Steere, R. L.: op. cit., p. 32.

Two-Phase Systems

Albertsson, P.: Two-phase Separation of Viruses, in *Methods in Virology,* K. Maramorosch and H. Koprowski (eds.), vol. II, pp. 303–321, Academic, New York, 1967.

CHAPTER 3 STRUCTURE OF VIRUSES

The first major development in the chemistry of viruses was Stanley's decisive demonstration in 1935 that crystalline tobacco mosaic virus is mainly protein. This was soon followed, as indicated in Chap. 1, by the discovery by Bawden and associates of ribonucleic acid in the highly purified virus. Thus it was shown that the major chemical constituents of TMV are protein and nucleic acid. This same general composition has been found over and over again for all kinds of viruses.

3-1 *Viral Composition*

VIRUSES ARE MAINLY NUCLEIC ACID AND PROTEIN

The generalization can now be made that all viruses in their mature form are composed minimally of one or more proteins and nucleic acid, except possibly for a few which may exist solely as nucleic acid (see Viroids, Chap. 7). However, some viruses are simple in chemical composition and others are more complex. Several features of the composition of viruses can be discerned from inspection of Table 3-1,

TABLE 3-1 **Approximate Composition of Some Viruses**

Virus	*RNA,* %	*DNA,* %	*Protein,** %	*Lipid,* %	*Non-nucleic acid carbohydrate,* %
Adeno (types 2 and 4)		13	87		
Alfalfa mosaic	19		81		
Avian myeloblastosis	2		62	35	1
Broad-bean mottle	22		78		
Broad-bean true mosaic	35		65		
Brome mosaic	21		79		
Carnation latent	6		94		
Cauliflower mosaic		16	84		
Coliphages f1, fd, M13		12	88		
Coliphages f2, fr, M12, MS2, R17, Qβ	30		70		
Coliphages ϕX174, ϕR, S13		26	74		
Coliphages T2, T4, T6		55	40		5†
Cowpea mosaic	33		67		
Cucumber mosaic	18		82		
Cucumber 3 (and 4)	5		95		
Dasychira pudibunda L.	7		93		
Encephalomyocarditis	30		70		
Equine abortion		9	+	+	
Equine encephalomyelitis	4		42	54	
Foot-and-mouth disease	31		69		
Fowl plague	2		68	25	+‡
Fowlpox		2	71	27	
Herpes simplex		9	67	22	2
Influenza	1		74	19	6
K (mouse)		7	93		
Mouse encephalitis (ME)	31		69		
Newcastle disease	1		73	20	6
Pea enation mosaic	29		71		
Poliomyelitis	26		74		
Polyoma		16	84		
Potato X	6		94		
Potato yellow dwarf	+		+	20	
Pseudomonas phage PM2		14	76	10	
Rabies	4		96		
Reo, type 3	21		79		+

TABLE 3-1 **(continued)**

Virus	*RNA,* %	*DNA,* %	*Protein,** %	*Lipid,* %	*Non-nucleic acid carbohydrate,* %
Rous sarcoma	2		62	35	1
Shope papilloma		18	82		
Silkworm jaundice		8	92		
Silkworm cytoplasmic polyhedrosis	23		77		
Sindbis (an arbor A virus)	9		47	44	
Southern bean mosaic	21		79		
SV5 (a simian parainfluenza virus)	1		73	20	6
SV40 (simian papova)		13	87		
Tipula iridescent		13	82	5	
Tobacco etch	5		95		
Tobacco mosaic	5		95		
Tobacco necrosis	20		80		
Tobacco rattle	5		95		
Tobacco ring spot	40		60		
Tomato bushy stunt	17		83		
Tomato spotted wilt	5		71	19	5
Turnip yellow mosaic	34		63		
Vaccinia		5	91	4	
Wild cucumber mosaic	35		65		
Wound tumor	23		77		

* Rounded figures, often obtained by finding the difference between 100 percent and the sum of other components.

† Hydroxymethylcytosine residues of the T-even DNAs are glucosylated but to different extents, so that glucose residues amount to about 4, 5, and 7 percent, respectively, of the DNAs of T2, T4, and T6. An average value for DNA and carbohydrate is given here.

‡ A + mark indicates that a component has been detected but its quantity has not yet been established.

from which it appears that the major constituents found in viruses are RNA or DNA, protein, lipid, and carbohydrate in addition to that which is a constituent of nucleic acid.

The occurrence and importance of each of these constituents will be briefly considered. First, let us turn our attention to nucleic acid, since this chemical constituent has been acclaimed the genetic stuff of all terrestrial life, including viruses.

3-2 *Viral Nucleic Acids*

VIRUSES CONTAIN EITHER RNA OR DNA BUT NOT BOTH. THEY ARE SINGLE-STRANDED, DOUBLE-STRANDED, AND CYCLIC POLYNUCLEOTIDES

Some viruses contain RNA and some DNA,[1] but none appears to contain both types of nucleic acid. In this respect viruses differ from bacteria, another major class of infectious disease agents (see Table 3-2). No unequivocal explanation for the occurrence of a specific type of nucleic acid (i.e., DNA or RNA) in a given virus has yet been proposed, although there must be some fundamental reason for this situation, perhaps related to the ultimate origin of a particular type of virus.

There is a considerable discrepancy in the proportion of nucleic acid found in different viruses. The range is from less than 1 percent of the mass of the virus particle (in influenza virus) to approximately half or more in coliphages T2, T4, and T6. Since nucleic acid is the hereditary material of a virus, the mass per particle (not shown in Table 3-1, but see Table 3-3) is doubtless proportional to the complexity of structure and number of functions of a given virus. Hence a complex virus requires more nucleic acid (many genes), and a simple virus less (few genes).

The simplest conceivable virus must consist of either RNA or DNA. In fact, however, currently known viruses when isolated from cells are almost universally more complex, the simplest, such as tobacco mosaic virus, being comprised of one species of protein and a single strand of nucleic acid. These two components, present in characteristic, constant proportions, are not covalently linked to one another, but generally achieve a stable and characteristic structure simply through intertwining of protein and nucleic acid and formation of secondary valence bonds between them. In the typical arrangement, the nucleic acid is found in an interior location sheltered from destruction by enzymes (nucleases). In order for this shelter to be

[1] The distinction between DNA and RNA can be made by qualitative analyses of the sugar and the purine and pyrimidine bases. The presence of deoxyribose and thymine indicates that the nucleic acid is DNA, whereas the detection of ribose and uracil indicates RNA.

TABLE 3-2 **Some Comparisons of Viruses and Bacteria**

Microorganism	*Size, nm (approx. diam.)*	*Chemical composition*	*Multiplication*	*Inhibition by antibiotics (e.g., penicillin, tetracyclines, sulfonamides)*	*Staining characteristics*
Bacteria	500–3,000	Complex: numerous proteins (including enzymes), carbohydrates, fats, etc.; DNA and RNA; cell wall contains mucopeptide	In fluids, artificial media, cell surfaces, or intracellularly, by binary fission	+	Stain with various dyes
Mycoplasma or PPLOs*	150–1,000	Similar to other bacteria but generally possessing no cell wall	In media similar to other bacteria but by budding rather than fission	Resistant to penicillins, sulfonamides; sensitive to tetracyclines and kanamycin	Stain with dyes but poorly
Rickettsia†	250–400	Similar to other bacteria	Inside living cells by binary fission; major hosts: *arthropods*	+	Stain with various dyes

Chlamydia or *Bedsonia*[†]	250–400	Similar to other bacteria	Inside living cells by binary fission; major hosts: *birds and mammals*	+	Stain with various dyes
Viruses	15–250	Mainly nucleic acid (one type) and protein. Some contain lipid and/or carbohydrate	Inside living cells by synthesis from pools of constituent chemicals	–	Stain for electron microscopy with salts of heavy metals

* PPLO is the abbreviation for pleuropneumonia-like organism, the first of this group of wall-less bacteria to be characterized.

† *Rickettsia* and *Chlamydia* are now considered to be small bacteria (see Stanier, Doudoroff, and Adelberg, *The Microbial World*, 3d ed., Prentice-Hall, Englewood Cliffs, N.J., 1970; Moulder, *The Psittacosis Group as Bacteria*, Wiley, New York, 1964; and Davis et al., *Microbiology*, Hoeber-Harper, New York, 1967).

TABLE 3-3 **Approximate Particle Weights and Nucleic Acid Contents of Some Viruses***

Virus	*Particle weight* $\times 10^{-6}$, *daltons*	*NA*† %	*NA*† *Type*	*NA per particle* $\times 10^{-6}$, *daltons*	*Approximate proportions nucleotides, Moles/100 moles*‡ *Ap*	*Gp*	*Cp*	*Up*	*dAp*	*dGp*	*dCp*	*dTp*	*dUp*	*5-HMdCp*	*5-HMdUp*
Adenovirus 2	177	13	DNA	23					21	29	29	21			
Avian myeloblastosis	500	2	RNA	10	25	29	23	23							
Bacillus subtilis PBS2									36	14	14		36		
Bacillus subtilis SP8	240	50	DNA	120					28	22	22				28
Broad-bean mottle	5	22	RNA	1	27	25	19	29							
Brome mosaic	5	21	RNA	1	27	28	21	24							
Coliphages f1, fd, M13	11	12	DNA-ss-c	1					24	20	21	35			
Coliphages f2, fr, M12, MS2, R17, β	4	30	RNA	1	23	26	26	25							
Coliphages ϕX174, ϕR, S13	6	26	DNA-ss-c	2					24	25	19	32			
Coliphage Qβ	4	28	RNA	1	22	24	25	29							
Coliphages T2, T4, T6	218	55	DNA	120					33	17		33		17	
Coliphage T3			DNA						25	25	25	25			
Coliphage lambda	57	56	DNA	32					26	24	24	26			
Coliphage T5			DNA	77					30	20	20	30			
Coliphage T7	50	47	DNA	24					26	24	24	26			

Cucumber 4	40	5	RNA	2	26	26	19	29				
Cytoplasmic polyhedrosis	55	23	RNA-ds	13	29	21	21	29				
Foot-and-mouth disease	7	31	RNA	2	26	24	28	22				
Fowlpox	10,000	2	DNA	200					32	18	18	32
Herpes simplex	900	9	DNA	81					16	34	34	16
Influenza	300	1	RNA	3	23	20	24	33				
Mouse encephalitis (ME)	6	31	RNA	2	25	24	24	27				
Newcastle disease	600	1	RNA	6	24	24	23	29				
Poliomyelitis	7	26	RNA	2	29	24	22	25				
Polyoma	22	16	DNA-c	4					26	24	24	26
Potato X	35	6	RNA	2	32	22	24	22				
Pseudomonas PM2		14	DNA									
Pseudorabies	1,100	5	DNA	55					14	36	36	14
Reo, type 3	70	21	RNA-ds	15	28	22	22	28				
Rous sarcoma	500	2	RNA	10	25	28	24	23				
Salmonella P22			DNA	28					25	25	25	25
Shope papilloma	29	18	DNA-c	5					26	24	24	26
Silkworm jaundice	276	8	DNA	22					30	20	20	30
Simian 40 (SV40)	40	13	DNA-c	5					26	24	24	26
Sindbis		6	RNA		29	26	25	20				
Tipula iridescent	1,200	13	DNA	156					34	16	16	34
Tobacco mosaic	40	5	RNA	2	28	24	22	28				

TABLE 3-3 **(continued)**

Virus	Particle weight $\times 10^{-6}$, daltons	NA[†] %	NA[†] Type	NA per particle $\times 10^{-6}$, daltons	Approximate proportions nucleotides, Moles/100 moles[‡] Ap	Gp	Cp	Up	dAp	dGp	dCp	dTp	dUp	5-HMdCp	5-HMdUp
Tobacco necrosis	8	20	RNA	2	28	26	22	26							
Tobacco necrosis satellite	2	20	RNA	0.4	28	25	22	25							
Tobacco ring spot	6	40	RNA	2	24	25	23	28							
Tomato bushy stunt	9	17	RNA	2	26	28	21	26							
Turnip yellow mosaic	6	34	RNA	2	23	17	38	22							
Vaccinia	3,200	5	DNA	160					30	20	20	30			
White clover mosaic	35	5	RNA	2	33	16	23	28							
Wild cucumber mosaic	7	35	RNA	3	18	16	40	26							
Wound tumor	70	23	RNA-ds	16	31	19	19	31							

* Values have been selected from the literature and rounded off to whole numbers. There is a wide variance in the accuracy of the values listed; hence they should be viewed in general as approximate.

† NA = nucleic acid. All RNAs are single-stranded (ss) unless otherwise noted, and all DNAs are double-stranded (ds) unless noted. Circular NA is denoted by c.

‡ AP = adenylic acid, dAp = deoxyadenylic acid, 5-HMdCp = 5-hydroxymethyldeoxycytidylic acid, etc.

effective, it is necessary that the protein be resistant to the proteolytic enzymes commonly found in cells, and this does in fact seem to be the case with most of the widely studied viruses. One can imagine that such structures were selected from several alternatives during the evolution of viruses.

The nucleic acids of viruses, like those found elsewhere in nature, appear to consist of unbranched polynucleotide chains. They can be released from virus particles by treating viruses with chemicals which disrupt secondary bonds (but not covalent linkages) and which unfold proteins. Phenol, sodium dodecyl sulfate, and guanidine hydrochloride are examples of reagents useful in effecting a separation of nucleic acid from other constituents of virus particles.

It will be noted that the nucleic acids of viruses listed in Table 3-3 are described as single-stranded, double-stranded, and cyclic. How is it known that these structures exist as described?

The current conception of the structure of nucleic acids, including viral nucleic acids, developed largely from the stimulating model put forth for DNA by Watson and Crick in 1953. This model started with the knowledge that DNA is composed of hundreds of nucleotides and hence is a polynucleotide structure. The nucleotides in the polymer are linked to one another through diester bonds between alternate sugar and phosphate groups, as shown in the sketches of Fig. 3-1. The type of structure depicted in Fig. 3-1 suggests that DNA should be a long, fibrillar molecule. Such threadlike structures, about 20 Å in diameter, were in fact observed in electron micrographs made of DNA by Williams and others between 1948 and 1953. The physical-chemical properties of DNA preparations were also consistent with a highly asymmetric molecule. For example, preparations of DNA were known to have extraordinarily high viscosities. Furthermore, since the viscosity calculated to infinite dilution seemed independent of ionic strength, the DNA molecule was judged to be rather more rigid than would be expected from the single-bonded character of the sugar-phosphate backbone. With respect to base composition of DNA, an outstanding but, in 1951, inexplicable feature emphasized by Chargaff was the observed molar equivalence of purines to pyrimidines.

Watson and Crick put the above facts together and added imaginatively the concepts of a helical structure and double-strandedness that were suggested by the x-ray diffraction studies of Wilkins, Franklin, and colleagues. The essential feature of the resulting model can be seen in the sketch shown in Fig. 3-2. This model shows two polynucleotide chains running in opposite directions in a helical con-

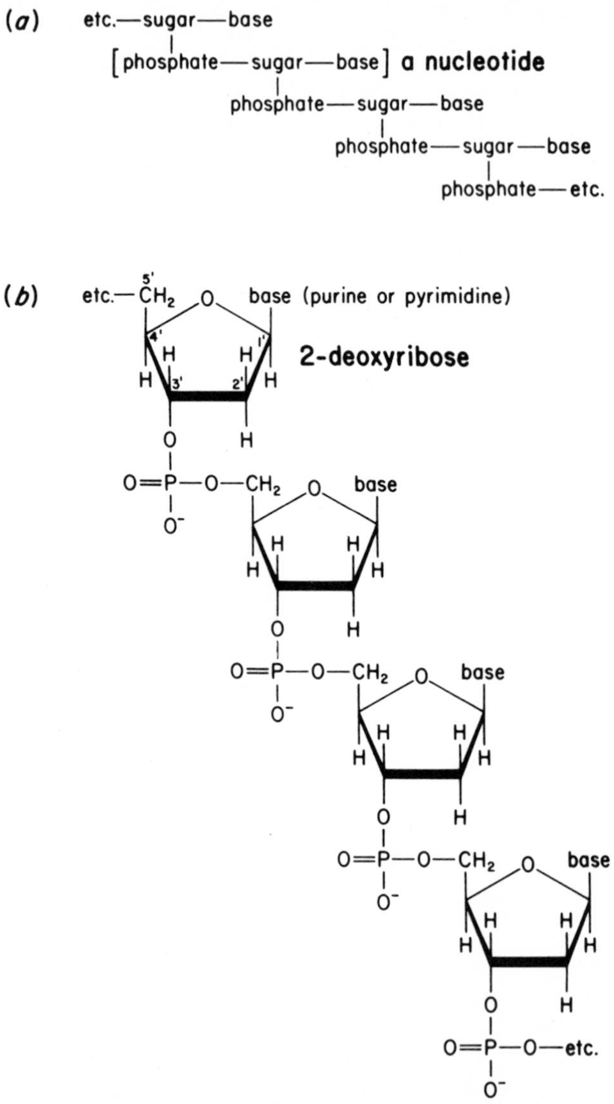

(*a*) etc.—sugar—base

[phosphate—sugar—base] a nucleotide

phosphate—sugar—base

phosphate—sugar—base

phosphate—etc.

(*b*) etc.—$\overset{5'}{C}H_2$ O base (purine or pyrimidine)

2-deoxyribose

$O{=}P{-}O{-}CH_2$ O base

O^-

$O{=}P{-}O{-}CH_2$ O base

O^-

$O{=}P{-}O{-}CH_2$ O base

O^-

$O{=}P{-}O{-}$etc.

O^-

FIG. 3-1. **(a) Composition of a very small segment of nucleic acid, indicating the order of arrangement of the three major components (sugar, base, and phosphate). Together they form a unit termed nucleotide. (b) Same as (a) but showing the bonding arrangement of the major constituents. In both sketches it can be noted that the structure is actually a chain composed of alternate sugar and phosphate links, each sugar link having either a purine or pyrimidine base attached to it. From the point of view of biologic synthesis, however, the structural unit is the nucleotide, i.e., phosphate-sugar-base, because the nucleotides, in the form of 5′-triphosphates, are in nature enzymatically joined to each other with release of phosphate to form the polynucleotide chain. By custom, the carbon atoms of the sugar moiety are designated with prime numbers, whereas the carbon and nitrogen atoms of the purine and pyrimidine bases are not.**

formation having a repeat distance along the fiber axis of 34 Å. (This figure is for the B structure which is the one assumed by DNA fibers with over 30 percent water content. A more compact form, the A structure, occurs when the water content is 30 percent. The two structures are interconvertible with change of moisture.) The two strands are held together by hydrogen bonds between the purine and pyrimidine bases, each base being linked to a companion base on the other strand. (Subsequently, evidence has been obtained that hydrophobic bonds contribute importantly to stability of the double helix.) The pairing of the bases is highly specific, so that guanine is almost always paired with cytosine and adenine with thymine, as shown in the sketch of Fig. 3-2. The hydrogen bonding between complementary base pairs occurs between hydrogen and nitrogen and hydrogen and oxygen, as illustrated in Fig. 3-3.

With this background on the structure of DNA, we may now return to the question of how one determines whether a viral nucleic acid is single-stranded, double-stranded, cyclic, or linear.

3-3 Primary Structure of Nucleic Acids

STRUCTURE OF A NUCLEIC ACID CAN BE DETERMINED BY MEANS OF ELECTRON MICROSCOPY, BASE ANALYSIS, MELTING BEHAVIOR, AND REACTIVITY TO CHEMICAL REAGENTS AND ENZYMES

Nucleic acids can be visualized with the electron microscope, especially if they are spread out in a protein monolayer by means of an ingenious technique devised by Kleinschmidt and his associates. The double-stranded DNA of Shope papilloma virus and the double-stranded RNA of reovirus are illustrated in Fig. 3-4. Single-stranded nucleic acid can be similarly visualized but is somewhat more difficult to distinguish from the background of the micrograph. It is easier to identify single-stranded nucleic acid if both single-stranded and double-stranded forms are present in the same field. Note, for example, the single-stranded filaments of RNA associated with the unusual circular piece of reovirus RNA in Fig. 3-4*b*. Cyclic nucleic acid can be detected nicely by electron microscopy and has been found to be characteristic of Shope rabbit papilloma virus, mouse polyoma virus, and simian virus 40 (sometimes these are called "papova" viruses).

In general, however, the strandedness of nucleic acids can best

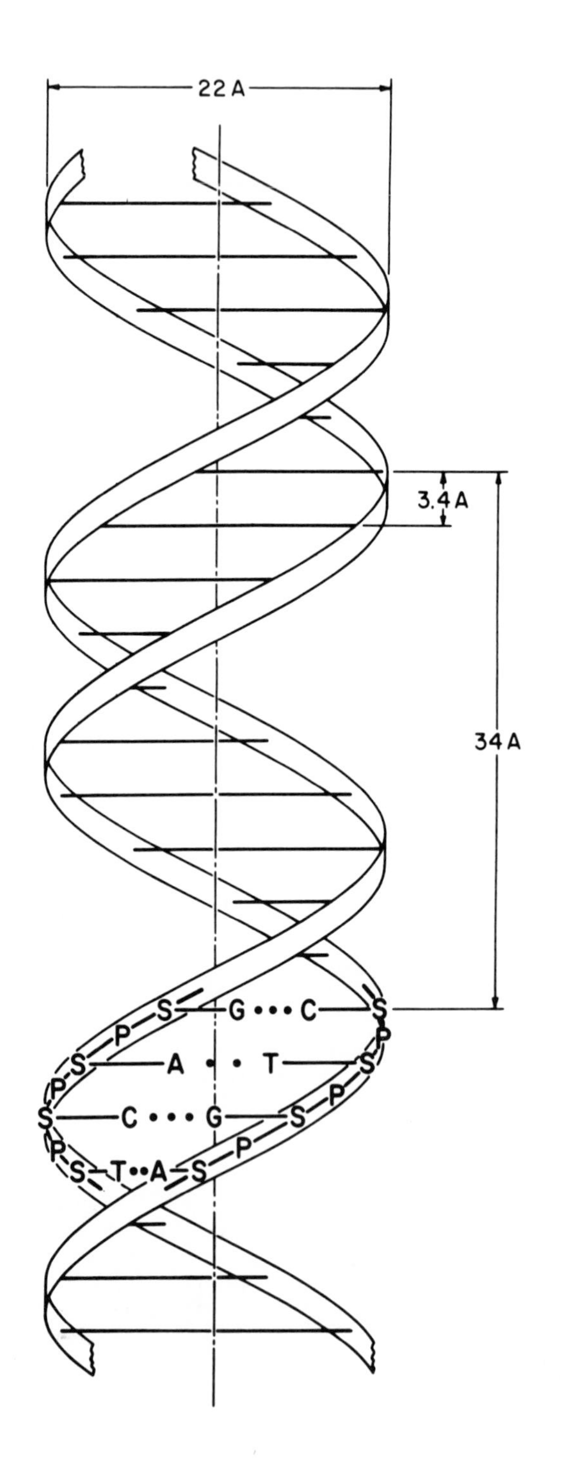

22 A
3.4 A
34 A
S—G•••C—S
S—A • • T—S
S—C•••G—S
P—S—T••A—S—P

be determined by chemical and physical measurements. For example, in double-stranded DNA the molar ratios of adenine to thymine and guanine to cytosine are both equal to unity, whereas these ratios deviate significantly from one in single-stranded DNA. Consequently, base analysis usually provides the means to distinguish between single-stranded and double-stranded nucleic acids. Data such as those given in Table 3-3 illustrate this point. Note, for example, the striking contrast between the nucleotide values for coliphage ϕX174 (which has single-stranded DNA) and vaccinia virus (which has double-stranded DNA). A similar comparison can be made with influenza virus and reovirus, which contain single-stranded RNA and double-stranded RNA, respectively. However, it should be noted that rare exceptions do occur in which molar equivalence of complementary bases can be observed with single-stranded nucleic acid. The single-stranded RNA of potato yellow dwarf virus represents such a fortuitous instance (guanine: 21 percent, cytosine: 21 percent, adenine: 29 percent, uracil: 29 percent).

Heating disrupts hydrogen bonds in nucleic acids whether they are between two strands, as in double-stranded nucleic acid, or between bases in loops of single-stranded nucleic acid. Such disruption of hydrogen bonds, called "molecular melting," can be detected more or less readily because it is accompanied by an increase in ultraviolet absorption at 260 nm (called "hyperchromic effect") and by decreases in viscosity and in optical rotation. The midpoint in the denaturation, as measured by absorbance change, is called the melting temperature, abbreviated T_m. The T_m for a synthetic, double-stranded polyadenylic-polythymidylic polymer is about 69°C, whereas in the same salt medium that of a double-stranded polyguanylic-polycytidylic polymer is about 104°C. This observation reflects the fact that there are two hydrogen bonds to be disrupted between

FIG. 3-2. **Diagrammatic sketch of the structure of DNA, modified from Watson and Crick, 1953, by indication of the components in a segment: P, phosphate; S, sugar (2-deoxyribose); G, guanine; C, cytosine; A, adenine; T, thymine. The two ribbons represent the sugar-phosphate backbones of the two helical strands of DNA, which, as the arrows indicate, run in opposite directions, each strand making a complete turn every 34 Å. The horizontal rods symbolize the paired purine and pyrimidine bases. There are 10 bases (and hence 10 nucleotides) on each strand per turn of the helix. The nature of the base pairs and the number of hydrogen bonds between them are shown in the detailed central segment. The vertical line marks the fiber axis.**

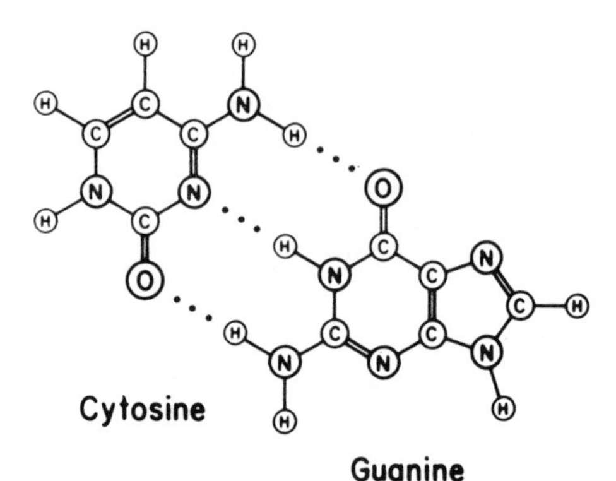

FIG. 3-3. **Pairs of purine and pyrimidine bases showing the kind of hydrogen bonding that occurs between them in nucleic acids.**

adenine and thymine, while there are three between guanine and cytosine, as shown in Fig. 3-3. Naturally occurring nucleic acids have T_m values falling somewhere between the extremes shown by the synthetic polymers cited above, and T_m values are proportional to guanine-cytosine content. Some examples of the melting curves for viral nucleic acids are shown in Fig. 3-5. As can be seen in the figure, there is a marked contrast in the types of curve obtained with single-stranded and double-stranded nucleic acids. In the case of the double-stranded nucleic acids, thermal disruption of structure, accompanied by increased absorbance in the ultraviolet, occurs within a narrow temperature range (sharp melting point) in the region of 80 to 105°C.

In contrast, the ultraviolet absorbency of partially hydrogen-bonded structures, such as single-stranded TMV-RNA, increases gradually as a function of temperature over a wide range (about 25 to 80°C).

Another distinction between single-stranded and double-stranded nucleic acids is in the reactivity of their purine and pyrimidine bases

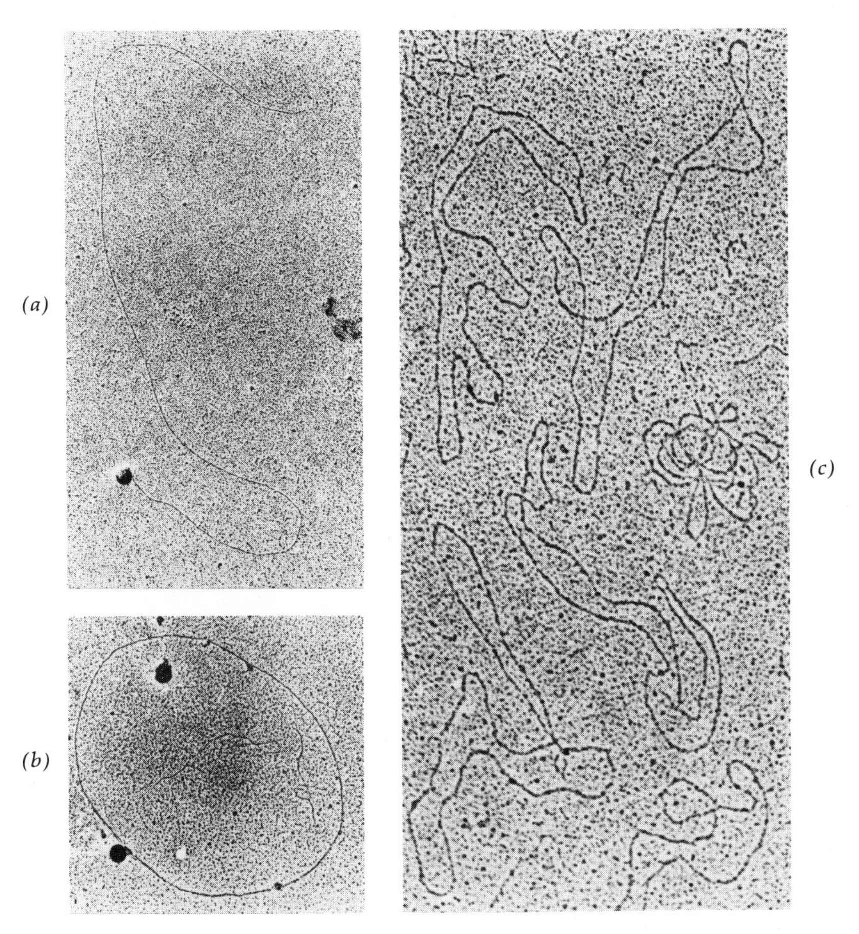

FIG. 3-4. **Electron micrographs of the double-stranded nucleic acids of reovirus and Shope papilloma virus. (a) and (b): Reovirus RNA. The nucleic acid commonly isolated from reovirus is linear and occurs in segments, but these remarkable filaments were observed by Dunnebacke and Kleinschmidt (Z. Naturforsch. 22b, 159–164, 1967) after short exposure of the virus to sodium perchlorate. (c): Shope papilloma virus DNA (Kleinschmidt et al., J. Mol. Biol. 13, 749–756, 1965).**

with chemical reagents. Double-stranded nucleic acids are less reactive because groups such as the amino groups of the purines and pyrimidines are heavily involved in hydrogen bridge formation and hence not readily available to chemical reagents. Thus the double-stranded form of phage ϕX174 DNA shows little evidence of reaction with formaldehyde, whereas the single-stranded DNA of this virus shows a typical response, namely, an increase in ultraviolet absorption and a displacement of the absorption maximum toward higher wavelengths (see Fig. 3-6).

The rate of degradation of nucleic acids, brought about by external agents such as hydrolytic enzymes or internally by decay of incorporated, radioactive phosphorus, serves also to distinguish between single- and double-stranded nucleic acids. With single-stranded molecules, any cleavage will at once reduce the average molecular weight (and infectivity), whereas with double-stranded nucleic acid, scission of the linear structure is not likely to occur at low temperature and moderate ionic strength unless cleavages are made in both

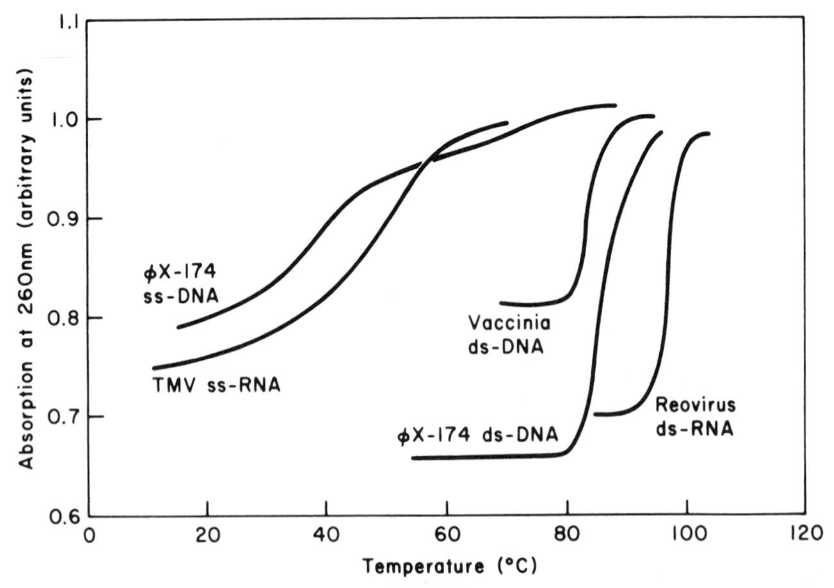

FIG. 3-5. **Variation of ultraviolet absorption with temperature of some viral nucleic acids in dilute salt solutions. The abbreviation ss stands for single-stranded, and ds for double-stranded. TMV is tobacco mosaic virus, and ϕX174 is a small coliphage. The ϕX174-dsDNA is a form observed in infected cells, whereas the ssDNA occurs in the mature phage particle.**

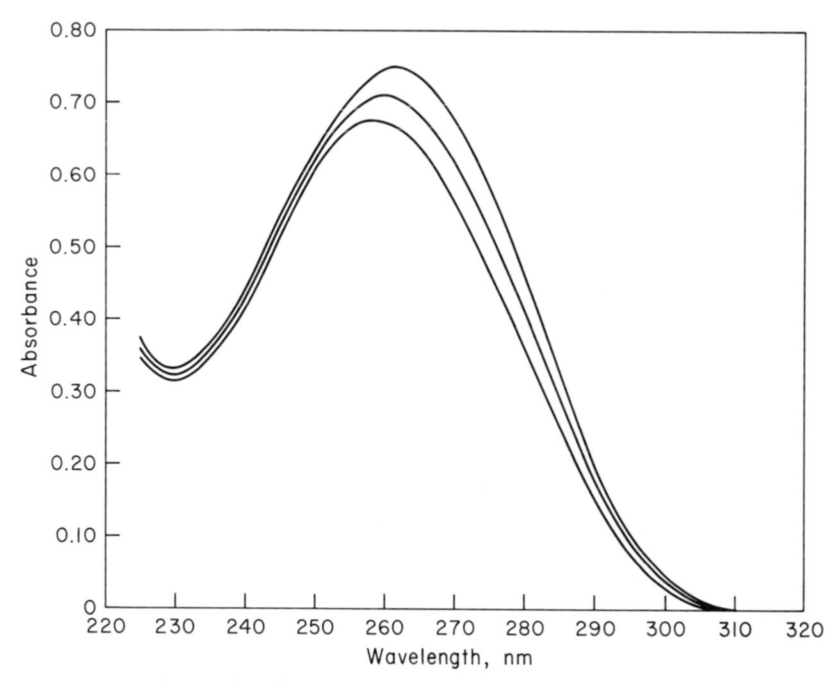

FIG. 3-6. **Ultraviolet absorption of single-stranded ϕX174-DNA at about 27 micrograms per ml in 0.2 M NaCl before and after treatment with 1.8 percent formaldehyde at 37°C. The bottom curve shows ultraviolet absorption of untreated ϕX-DNA, the middle curve shows absorption after exposure to formaldehyde for 36 min, and the top curve after 6 to 18 hr. Note the increase in absorption after exposure to formaldehyde and the displacement of the maximum from about 260 to about 264 nm. (From Sinsheimer, J. Mol. Biol. 1:43–53, 1959.)**

strands and in the same vicinity. The occurrence of breaks in nucleic acids can be detected by such means as viscosity measurements, light scattering, or sedimentation behavior.

3-4 *Nucleic Acid Function*

NUCLEIC ACIDS CONSTITUTE THE GENETIC MATERIAL OF NATURE

Several discoveries contributed substantially to the recognition of nucleic acids as the genetic material of nature. One of these involved bacteria and the others viruses, as follows:

1. DNA Causes Bacterial Transformation. Avery, MacLeod, and McCarty reported in 1944 that pneumococcus type II, characterized by unencapsulated cells, could be transformed into pneumococcus type III with typical encapsulated cells by treatment of type II cells with DNA extracted from type III. Since the presence or absence of capsules was known to be a hereditary trait, it was concluded that DNA was the hereditary chemical of pneumococci. In the words of Avery and his collaborators, "The evidence presented supports the belief that a nucleic acid of the desoxyribose type is the fundamental unit of the transforming principle of Pneumococcus Type III."

2. Analyses of Spontaneous Mutants of Tobacco Mosaic Virus Suggest a Genetic Role for RNA. A finding of little influence at the time, despite its suggestive nature, was the report by Knight in 1947 that some spontaneous mutants of TMV which evoked different symptoms in tobacco apparently had identical protein components. By default, the RNA would have to be, as suggested, the genetic component of the virus. In addition, it was found in these same experiments that the proteins of some other mutant strains of TMV differed from that of the common type in content of certain amino acids but not in others. It was difficult to visualize how such chemical changes could have been effected as a primary mutational event. Hence the situation was summarized by the statement: "It is not known as yet whether the chemical differences observed between virus strains are the primary results of mutation or whether they are the secondary results of critical changes which have taken place first in, for example, the nucleic acid components."

3. Particles of Turnip Yellow Mosaic Virus Lacking Nucleic Acid Are Not Infectious. In 1949 Markham and Smith reported the isolation and crystallization of turnip yellow mosaic virus. When the sedimentation of such highly purified preparations of this virus was studied it was found that instead of the expected single species of particle there were two. These were separated on the basis of their different sedimentation rates, and their properties were compared. The two kinds of particles proved to be virtually identical in several properties such as size, shape, crystalline form, electrophoretic mobility, and serologic reaction. The crucial difference, which was responsible for the different sedimentation rates observed, was that the particles comprising the faster-sedimenting population contained about 37 percent RNA, whereas the slower-sedimenting particles had little or no RNA. Significantly, the particles containing RNA were found to be highly infectious, while those lacking RNA were not infectious.

Two interpretations of these results were apparent: both nucleic acid and protein are required for infectious capacity, or nucleic acid is the genetic component of the virus. Markham took the latter view when he concluded in a paper presented at Oxford in 1952: "The role of the protein constituent of plant viruses is undoubtedly very important, but there is some evidence that the nucleic acid is in fact the substance directly controlling virus multiplication."

It should be noted that the experiments on transformation of bacteria pointed to DNA as the genetic material of these microorganisms, whereas the work on tobacco mosaic virus and on turnip yellow mosaic virus indicated that RNA may also have a genetic function.

4. Infection of Bacteria by Some Coliphages Results from Injection of Viral DNA into the Bacterial Cells. The use of radioisotopes (^{35}S and ^{32}P) to tag the protein and nucleic acid components, respectively, of coliphage T2 enabled Hershey and Chase in 1952 to conclude that DNA is the genetic material of the T phages. Specifically, it was shown that at least 80 percent of the viral sulfur, which marked the phage protein, remained outside infected cells and could subsequently be shaken off without altering the course of infection. Conversely, the bulk of the radioactive phosphorus, marking the nucleic acid, became closely associated with the infected cells and presumably penetrated them, since a considerable portion could be recovered from the phage progeny issuing as a result of the infection. Very little (less than 1 percent) of the infecting phage sulfur appeared in progeny. These and other facts led to the summary: "The sulfur-containing protein of resting phage particles is confined to a protective coat that is responsible for the adsorption to bacteria, and functions as an instrument for the injection of the phage DNA into the cell. This protein probably has no function in the growth of intracellular phage. The DNA has some function."

5. The Genetic Role of RNA from Tobacco Mosaic Virus Is Directly Demonstrable. It was found by Gierer and Schramm in 1956 that treatment of TMV with aqueous phenol resulted in the release of infectious RNA into the water layer. At about the same time, infectious TMV-RNA preparations were reported by Fraenkel-Conrat, who obtained his material by disaggregating the virus with the detergent, sodium dodecyl sulfate. Infection of plants with such RNA preparations resulted in the production of symptoms exactly like those produced as a consequence of infection with whole virus. Moreover, progeny virus particles obtained from such infections were complete, with their characteristic protein coats. It was concluded that infectivity, with all its attendant hereditary characteristics, was due to the RNA.

All the major premises of these early investigations have been substantiated and extended by subsequent experiments in a wide variety of systems. Consequently, the concept of nucleic acids as the genetic substances of nature is now largely taken for granted. In this scheme, DNA is the main hereditary material for man, other animals, plants, and microorganisms, including many viruses. RNA serves this function for many other viruses.

3-5 Primary Structure of Nucleic Acids

DETAILED INFORMATION ABOUT VIRAL NUCLEIC ACIDS IS OBTAINED BY DETERMINING SEQUENCES OF NUCLEOTIDES

Specific sequences of amino acids in proteins originate from corresponding specific sequences of nucleotides in nucleic acids (the genetic code is given in Table 6-3). Therefore, a prime objective of molecular genetics is to determine the sequence of nucleotides in various nucleic acids. This is a very difficult analytic problem. Since the sequencing of nucleotides in a nucleic acid and the sequencing of amino acids in a protein are similar problems, it may be useful to compare the magnitude of the two endeavors. First, if determinations of complete sequences are desired, the total number of amino acids in a viral protein ranges from about 49 to 2,000, whereas the smallest viral nucleic acid has around 3,200 nucleotides and the largest, several hundred thousand. Clearly much more sequencing is required to determine the complete primary structure of a viral nucleic acid than to do the same for a viral protein. Next, a viral protein is composed of 16 to 20 different amino acids, whereas a viral nucleic acid contains only four different nucleotides. The combination of greater numbers of structural units (nucleotides) but fewer different types results in fewer distinctive segments of nucleic acid than in the case for proteins. This is important because no methods are available for complete sequencing of either polypeptide or polynucleotide chains starting from one end and progressing through to the other end. In both cases, the chains of structural units are broken into small segments and these are sequenced. Ultimately the individual sequences have to be pieced together to give a total sequence, and here it is crucial to have segments which can be distinguished from one another. Such short novel sequences tend to be more common in proteins than in nucleic acids. A third possible complication in sequencing nucleic acids which is lacking with proteins is that some nucleic acids are double-stranded. In such cases it is necessary to separate the strands cleanly before beginning the sequencing. This may be difficult.

Nevertheless, using techniques developed by F. Sanger and colleagues, substantial progress has been made in sequencing some of the smaller viral nucleic acids, especially those of the RNA phages. The general procedure employed is to subject the isolated RNA to limited digestion by a specific nuclease enzyme. The digestion products, oligonucleotides containing up to 50 nucleotides, are fractionated by a two-dimensional procedure using electrophoresis (ionophoresis) on cellulose acetate in the first dimension and ascending chromatography on thin layers of DEAE-cellulose and cellulose in the second dimension. Oligonucleotide spots are located by autoradiography, eluted, and subjected to further digestion with enzymes, and these products are separated by the two-dimensional method. Analyses for the nucleotides present in eluted spots, together with application of some stepwise degradation procedures, lead ultimately to a series of partial sequences. A few examples from analyses of the RNAs of R17 and f2 phages are given in Table 3-4, showing some homologous oligonucleotides with small differences and one pair of identical segments.

TABLE 3-4 **Sequences of Some Homologous Oligonucleotides from the Ribonucleic Acids of R17 and f2 Phages**

Oligonucleotide from	*Sequence*	*Differences*
R17	A-A-U-U-A-A-C-U-A-U-U-C-C-A-A-U-U-U-U-C-Gp	U → C
f2*	—— C————————————————————	
R17	C-U-A-U-A-U-U-C-A-U-A-U-C-U-U-Gp	2U → 2C
f2	———————————————— C—(U,C) Gp	
R17		
f2	A-A-A-U-U-U-A-C-C-A-A-U-C-A-A-U-U-Gp	None
R17	U-U-U-U-A-C-A-A-A-C-C-A-Gp	C → A
f2	——————————————A————	

* A line for f2 sequence signifies that it is the same as R17. Sequences given in parentheses indicate uncertainty of order of the parenthetical nucleotides.

SOURCE: Robertson and Jeppesen, 1972.

3-6 *Gene Location*

VIRAL GENES ARE SPECIFIC SEGMENTS OF NUCLEIC ACID CONTAINING 150 TO 1,000 OR MORE NUCLEOTIDES, DEPENDING UPON THE GENE CONCERNED; VIRAL GENES HAVE BEEN LOCATED BY MATING EXPERIMENTS, ELECTRON MICROSCOPY, SELECTIVE MUTAGENESIS, AND SEQUENCING OF NUCLEOTIDES

Mating is used as a tool for mapping phage genes. The principle is that phage genes like those in the chromosomes of plants and animals are linked in a specific linear sequence and that the frequency with which genes in homologous parts of different chromosomes are exchanged is proportional to the distance separating the genes involved. Consequently, if suitable markers are available as distinctive hereditary characters, crosses can be made with a series of strains, and from the results the genes associated with the hereditary characters can be mapped. A simple illustration can be derived from mixed infection data obtained with some plaque-type mutants of coliphage lambda and listed in Table 3-5.

As shown in Table 3-5, four products were observed as a conse-

TABLE 3-5 **Recombination between Mutants of Coliphage Lambda**

*Mutants crossed**	*Progeny, nos. of plaques*				*Recombination %†*
Rc × *Rm*	5,162 *Rc*	6,510 *Rm*	311 *RR*	341 *cm*	5.3
Rs × *Rc*	7,101 *Rs*	5,851 *Rc*	145 *RR*	169 *cs*	2.4
Rs × *Rm*	647 *Rs*	502 *Rm*	65 *RR*	56 *sm*	9.5

* Adapted from Kaiser (*Virology* 1, 424, 1955). The reference type of phage in these studies was a variant of common (wild type) phage lambda which appeared as an exceptional plaque upon plating a lysate of *Escherichia coli* K12 on *E. coli* strain C600. This reference phage, which will be designated *RR* here, was characterized by plaques which were uniform, large, and turbid. From the reference phage several mutants were obtained by ultraviolet irradiation. The ones used in mixed infections with the results shown here were, according to type of plaque they produced when plated on *E. coli* strain C600: *Rs*, small turbid plaques; *Rm*, minute plaques (smaller than *s*); *Rc*, essentially clear plaques about the same size as those of the reference phage.

† Recombination percent is obtained from the relation:

$$\frac{\text{Sum of two recombinants}}{\text{Sum of all types}} \times 100$$

quence of each cross. Two of the products were the parental phage types, one was reference type, and the fourth was a double mutant. A linkage map showing the relative positions of the mutant plaque genes can be constructed from the data of Table 3-5. A common procedure, which is used here, is to set each percent of recombination equal to one map unit. Thus the *c* and *m* genes are separated by 5.3 map units, and the *s* and *c* genes by 2.4 units. However, it is not possible to decide on which side of *c* the *s* gene is located until the third cross is made, *Rs* × *Rm*. The results of this cross (Table 3-5) indicate that *s* must be on the distal side of *c* rather than between *c* and *m*, thus giving the linkage arrangement:

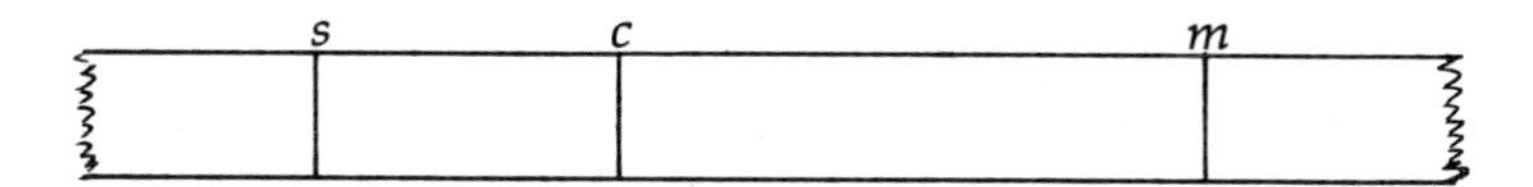

A striking departure from the linear map was observed for phage T4, and later for certain other phages, by Streisinger and associates. The essence of the anomalous finding was that in a three-factor cross a particular gene locus failed to segregate independently of two other presumably remote loci, and furthermore the results indicated that this locus was closer to the more distant (as deduced from other crosses) of the two remote loci than to the nearer locus. This result could be explained on the basis of a circular genetic map. However, even though the DNAs of some phages and of *Escherichia coli* have a circular structure (Table 3-3), this is not the case for the DNAs of the T-even phages. Therefore, the apparent circularity of such genomes as that of T4 phage must be explained in some other manner. A popular concept, which also has experimental backing, is that the genes in T4 and similar phages are circularly permuted and a few genes at one end of the linear genome are repeated at the other end (terminal redundancy). This is illustrated in Table 3-6.

TABLE 3-6 **Illustration of Circularly Permuted Gene Order and Terminal Redundancy in Three Linear, Homologous Genomes**

1	2	3	4	5	6	7	8	9	10	1
2	3	4	5	6	7	8	9	10	1	2
3	4	5	6	7	8	9	10	1	2	3

Thus it is consistent with many experimental findings to represent the linkage of genes in a virus such as T4 with a circular map. Such a map, showing also some of the functions associated with various genes, is illustrated by Fig. 3-7.

It can be noted from inspection of the T4 map that the genes are not randomly arranged in the linkage but tend to cluster with respect to function. Hence many of the genes concerned with phage head are clustered together, while those involved with tail fibers are mostly in another cluster. Likewise, several of the genes that determine early phage functions, such as synthesis of viral DNA, are grouped together, and those concerned with assembly of the virus particle are in another group. There are exceptions, but the trend is apparent, even though the mechanistic significance of this arrangement is not yet clear. Clusters of genes of like function have also been observed in the mapping of animal virus genomes.

If the viral nucleic acid is double-stranded and mutants are available, it is possible to map some genes visually by a combination of electron microscopy and hybridization techniques. The principle of the technique as applied to location of a deletion in phage-lambda DNA may be briefly summarized as follows. Purified concentrated

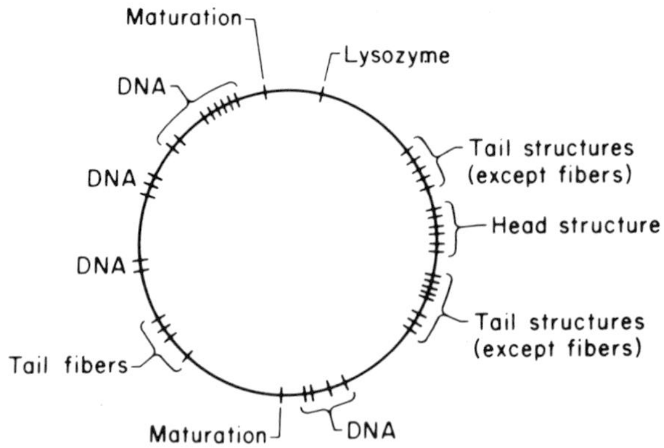

FIG. 3-7. **Approximate map location of some of the genes of bacteriophage T4. The genes are designated by the names of viral components whose production and assembly are controlled by the genes at the loci indicated. Two other genes concerned with completion of assembly of infectious phage particles and called "maturation genes" are also shown. (Adapted from Edgar and Wood, 1966, with omission of many genes in order to show more clearly the clustering of genes of like function.)**

stocks of wild type phage lambda and of a deletion mutant are subjected to a brief heat treatment in the presence of a detergent. This releases the DNA from the phage particle and causes separation of the two complementary strands of the DNA. The separated strands of DNA are obtained in different fractions upon density-gradient centrifugation under special conditions (such as in the presence of poly I, G). A strand from the deletion mutant can then be annealed with the complementary strand from wild type. Base pairing occurs regularly between the strands except in the deleted region of the mutant. Since there is nothing to pair with in this region, the unpaired portion of the wild type strand loops out, and pairing continues regularly beyond the loop-out section. The loop-out region, and hence the section deleted, can be precisely located visually by Kleinschmidt's very useful technique of spreading the nucleic acid in a protein monolayer, followed by electron microscopy. The kind of result obtained is shown diagrammatically in Fig. 3-8.

Another example of the hybridization–electron microscopy technique for gene location is its use by Morrow and Berg for the identification of a small segment of simian virus 40 (SV40) genome in the hybrid species containing both adenovirus and SV40 DNAs. A mixture of the DNAs from the hybrid species (called $Ad2^+ND_1$, which was one of the products resulting from mixed infection of monkey kidney cells with adenovirus 2 and SV40) and adenovirus-2 DNA were denatured to their single-stranded components in alkaline solution and renatured to double-stranded structures in formamide. Among the products was found a mixed double-stranded species (called a *heteroduplex*) which consisted of one strand from adenovirus 2 (Ad2) and a complementary strand from the hybrid, $Ad2^+ND_1$. Since part of the Ad2 DNA had been replaced by a segment of SV40 DNA in the hybrid, a portion of the hybrid DNA could not anneal to the Ad2 DNA, and hence a single-stranded loop-out of the Ad2 DNA opposite the SV40 segment was visible in the electron micrograph of the heteroduplex. In order to identify the precise segment of SV40 DNA in the hybrid, a triple heteroduplex was formed by denaturing a mixture of SV40 DNA (whose ring structure was first opened at a specific site by treatment with an *E. coli* endonuclease enzyme), hybrid DNA, and Ad2 DNA and then annealing the resulting strands. A heteroduplex between Ad2 and hybrid DNAs was obtained as before, but in addition, annealing occurred between the looped-out segment of SV40 DNA of the hybrid and its complementary region in the strand of DNA derived from complete SV40 DNA. Thus by inspection of the electron micrographs and by measuring distances from the endonuclease cleavage site to the point where

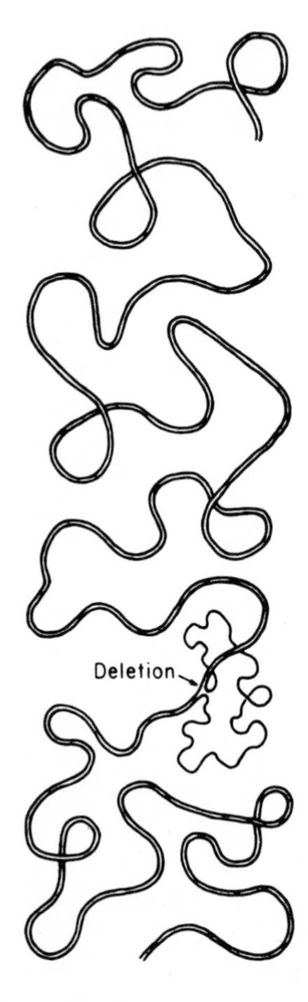

FIG. 3-8. **Schematic representation of the structure obtained upon annealing one strand of DNA from wild type phage with a complementary strand of DNA from a deletion mutant of that phage. The position on the mutant strand of DNA from which a segment has been deleted is noted, and opposite it the corresponding looped-out segment of the wild type strand. (Adapted from an electron micrograph of such a situation published by Westmoreland et al., 1969.)**

SV40 DNA participated in the formation of triple heteroduplex, it was possible to identify the segment of SV40 DNA containing the genes concerned with such functions as production of a specific SV40 antigen and the capacity to multiply in monkey kidney cells (Ad2 lacks this capacity).

It has not been possible as yet to map genes of plant viruses by mating experiments with viral mutants. Nor has the electron-microscope method for detecting deletion mutants been applicable. However, a novel method of mapping genes in tobacco mosaic virus RNA was developed by May, Kado, and Knight. The procedure takes advantage of the linear nature of the TMV particle and is based on the removal of coat-protein subunits from the rodlike particle in a highly specific manner, starting from the 3′ end of the RNA. This is done by treating the virus with sodium dodecyl sulfate at 37°C for about 3 hr.

Such treatment provides large quantities of particles which are stripped from 0 to about 80 percent, with a peak in concentration of particles that are 50 to 60 percent stripped. This population of particles is fractionated by sedimentation on a sucrose gradient to give classes of particles stripped to various degrees. Such classes of partially stripped particles are next treated with a mutagenic reagent, nitrous acid, and then tested for the mutagenic effect of the nitrous acid on a particular genetic marker. For example, wild type TMV gives a systemic green-mottling disease in a kind of tobacco called *Nicotiana sylvestris,* but a certain class of mutants readily produced by treating the virus or its RNA with nitrous acid causes local lesions (brown necrotic spots) on the leaves of sylvestris. Classes of TMV, partially stripped of protein subunits by treatment with dodecyl sulfate as sketched above, were treated with nitrous acid and then tested on sylvestris tobacco. A substantial increase in mutant lesions was observed only after the particles were about 70 percent stripped (Table 3-7). It appears from these results that the gene concerned in the local-lesion response on sylvestris tobacco is located in the first quarter of the 5′ end of the TMV-RNA.

Yet another set of procedures has been employed to determine that there are three genes in the RNA of phage R17 and to fix the order in which these genes occur. It appears that the RNAs of coliphages of the R17 series (i.e., R17, MS2, fr, and f2) code for only three proteins: the main coat protein, A protein (or maturation protein), and a specific RNA synthetase responsible for replicating the phage RNA. Each of these proteins has been isolated, and a complete or partial sequence of its amino acids has been determined. Either whole phage RNA or parts can direct synthesis of phage proteins

TABLE 3-7 **Relation of Mutagenesis by Nitrous Acid to Degree of Stripping of Protein Subunits from the RNA Genome of Tobacco Mosaic Virus**

Orientation of TMV-RNA	*Degree of stripping, %*	*Local-lesion mutants per 1,000 infectious units*
3′	0–20	0.2
↓	51–60	1.2
↓	61–70	3.3
↓	71–80	7.2
5′	100	10

in vitro. Treatment of phage RNA with *E. coli* ribonuclease IV cleaves the molecule into two pieces, one comprising about 40 percent of the RNA from the 5′ end and the remaining 60 percent containing the original 3′ end of the RNA. When these fragments of RNA were tested for capacity to direct protein synthesis in vitro in the presence of *E. coli* ribosomes, it was found that the 60 percent fragment is translated to yield synthetase protein, while the 40 percent fragment is inactive as messenger. When ribosomes from *Bacillus stearothermophilus* are used in the in vitro synthesis in place of *E. coli* ribosomes, only the A protein appears to be synthesized, and this is directed by the 40 percent fragment of RNA. These results suggest that the gene for A protein is near the 5′ end of the phage RNA and that the gene for synthetase is near the 3′ end. By difference this would leave the coat-protein gene in the middle.

Further evidence for this orientation was obtained as follows. Ribosomes from *E. coli* combine with and shield from nuclease digestion the regions of RNA near the beginning of the three genes. These are called "initiation regions," since protein synthesis starts at these sites. The 40 and 60 percent fragments of R17-RNA described above bind to *E. coli* ribosomes as noted earlier, but instead of allowing the fragments to be translated in vitro as before, the unbound, exposed pieces of RNA are digested off with ribonuclease. The RNA initiation regions containing fragments of their adjacent genes can then be released from the ribosomes by treatment with 8 *M* urea and purified by passage through a small DEAE-cellulose column. The isolated RNA initiation regions were subjected to nucleotide sequencing by the methods outlined earlier. The results obtained indicate that some of the oligonucleotides isolated represent portions of the genes of each of the R17 proteins together with nucleotides presumed to be part of ribosome binding sites which specify initiation of protein synthesis. These nucleotide sequences, together with the corresponding amino acids for which they code and which match the known N-terminal amino acid sequences of the three R17 proteins, are shown in Table 3-8. Not shown in the table are nucleotide sequences which by comparison with known amino acid sequences of the proteins were recognized as parts of the coat-protein gene. One oligonucleotide thus identified with the N-terminal portion of coat protein was found in the earlier mentioned 40 percent fragment of R17-RNA, whereas two other oligonucleotides corresponding to the interior of the coat protein were found in the 60 percent piece. This links the coat-protein gene to the central region of the RNA. Therefore, the order of genes in the R17-RNA going from the 5′ to the 3′ end of the molecule is A protein–coat protein–synthetase.

3-7 *Viral Proteins*

VIRUSES HAVE ONE OR MORE PROTEINS, AND OFTEN EACH OF THESE IS COMPOSED OF MANY IDENTICAL SUBUNITS

The 20 common amino acids found in nature also occur in viral structural proteins (and in virus-induced enzymes), with little or no indication of the presence of unusual amino acids. Nor is there evidence for extraordinary quantities of any amino acid in most viral proteins. One might have thought that there would be a preponderance of basic amino acids in the viral proteins most closely associated with the viral nucleic acid, but this seems not to be the case. In fact, many viral proteins have proved to be acidic, the isoelectric point of the common strain of tobacco mosaic virus being as low as pH 3.5.

Viral structural proteins are obtained from purified viruses by treatments that dissociate proteins from nucleic acids and from other proteins without breaking peptide bonds. This can often be done by treatment of the virus with dilute alkali or acid or with 1 or 2 *M* salts of various kinds. Phenol or sodium dodecyl sulfate can also be used, but the first reagent tends to denature proteins excessively, and the second binds almost irreversibly. In general, the amount of irreversible denaturation may be assumed to be small if the isolated protein is soluble in dilute salt at neutral or slightly alkaline pH values, and if the protein binds about the same quantity of specific antibody as it did when incorporated in the virus particles.

TABLE 3-8 **Nucleotide Sequence of Initiation-Site Oligonucleotides Isolated from R17-RNA***

	Oligonucleotide	*Observed N-terminal amino acid sequence corresponds to:*
	f-met arg ala phe ser	
A	AUUCCUAGGAGGUUUGACCUAUGCGAGCUUUUAGU	A protein
	f-met ala ser asn phe	
C	AGAGCCCUCAACCGGGGUUUGAAGCAUGGCUUCUAACUUU	Coat protein
	f-met ser lys thr thr lys	
S	AAACAUGAGGAUUACCCAUGUCGAAGACAACAAAG	Synthetase

* The nucleotides on the left of each oligonucleotide chain are considered to be parts of the initiation regions to which ribosomal units bind preceding translation.

Beginning in about 1937 and continuing for some time thereafter, Pirie, Stanley, Schramm, and their respective associates studied the degradation of plant viruses by such chemicals as urea, alkali, and detergents. In another approach, Bernal and Fankuchen made low-angle x-ray diffraction studies on oriented gels of plant viruses. The results of both these types of investigation suggested that virus structural proteins are made up of subunits of relatively low molecular weight. However, this idea remained rather vague until Harris and Knight showed by means of protein end-group studies that TMV has hundreds of equivalent peptide chains (protein subunits) having a molecular weight of about 17,000 and terminating in the amino acid threonine. The concept of viral protein subunits was subsequently generalized and popularized by Crick and Watson. It is now taken for granted that whether a virus has a single species of structural protein or 10, each protein will probably be found to consist of identical subunits, the number and composition of which serve to distinguish it from the other proteins. (An exception is simian virus 40, whose coat protein appears to consist of three distinct species of protein of similar molecular weight.) This arrangement is obviously economical with respect to the amount of nucleic acid needed to code for the viral proteins. On the other hand, it poses some conceptual problems concerning the mechanics of assembly (especially of those viruses with several protein components), and with respect to stability of viral structures, since the subunits are not covalently linked. Nature, of course, solves these problems quite readily, for virus particles of considerable perfection and stability are produced in abundance. Some examples of the numbers and sizes of viral protein subunits are given in Table 3-9.

3-8 Protein Function

VIRAL STRUCTURAL PROTEIN DETERMINES PARTICLE STRUCTURE, SPECIFICITY OF ATTACHMENT TO CELL RECEPTOR SITES, MUCH OF THE SEROLOGIC SPECIFICITY, AND SOME ENZYMATIC ACTIVITIES, AND AFFORDS PROTECTION TO VIRAL NUCLEIC ACID

As indicated in Table 3-1, protein constitutes most of the mass of viruses of all sorts. It is not surprising, therefore, that the size and shape of virus particles are largely determined by their constituent protein or proteins. The configuration of protein molecules is in turn determined by the sequence of their constituent amino acids, and these sequences are dictated by the nucleotide sequences (genetic code) of the viral nucleic acid.

The first step in infection by many viruses is a specific adsorption or attachment of the virus particle to a receptor site on the surface of a susceptible cell.

TABLE 3-9 **The Protein Subunits of Some Viruses**

Virus	*No. of constituent proteins*	*Approximate molecular weight of subunits*
Alfalfa (lucerne) mosaic	1	25,000
Broad-bean mottle	1	20,500
Brome-grass mosaic	1	18,000
Cucumber 4	1	15,000
Potato X	1	27,000
Shope papilloma	1	40,000
Southern bean mosaic	1	30,000
Sowbane mosaic	1	19,000
Tobacco mosaic	1	17,500
Tobacco necrosis	1	30,000
Tobacco necrosis satellite	1	20,000
Tobacco rattle	1	24,000
Tomato bushy stunt	1	41,000
Turnip yellow mosaic	1	20,000
White clover mosaic	1	22,500
Adeno	9	7,500–120,000
Coliphage f2	2	14,000 and 39,000
Coliphage X174	4	5,000–48,000
Coliphage T4	28	11,000–120,000
Herpes simplex	ca. 24	25,000–275,000
Influenza	ca. 7	25,000–94,000
Newcastle disease	ca. 6	38,000–74,000
Polio	4	6,000–35,000
Polyoma	6	13,000–43,000
Reovirus	ca. 7	34,000–155,000
Simian 40	3	16,000–17,000
Vaccinia	ca. 30	8,000–200,000

In the case of the tailed bacteriophages, the attachment specificity on the virus side resides in the protein of the tail fibers of the phage particle. Phages whose tail fiber proteins do not have affinity for any receptor sites on bacterial cell surfaces will not be able to attach to and hence infect those cells. The spheroidal, RNA-containing phages of the R17-f2-fr series contain, in addition to a coat protein which occurs in the many identical subunits, a special protein (probably one molecule per virus particle) which is essential for infectivity. This protein, called A protein, is apparently involved directly or indirectly (it may modify the conformation of the coat protein) in attachment to a susceptible cell. Steitz isolated the A protein and showed that it is easily distinguished from the coat protein because it is considerably larger (about 38,000 versus 14,000 M.W.) and it contains five histidine residues, whereas the coat protein has none. There is some indication that Rous sarcoma virus may also have a protein component important in maturation and attachment phenomena. Attachment specificity is dramatically demonstrated with poliovirus, whose coat protein enables it to attach to and infect only primate cells, whereas the RNA isolated from poliovirus can infect cells of many species. Thus the host specificity of animal and bacterial viruses depends primarily on the structure of one or more of the proteins of the virus.

The macromolecules found in nature, including those comprising virus particles, such as proteins, carbohydrates, nucleic acids, and lipids, are all antigenic, which means that they are capable of eliciting the formation of antibodies when injected into appropriate animals. Antibodies combine specifically with the substances eliciting their formation. Such combination is often sufficient to abolish the infectivity, i.e., to neutralize the virus. The protein components of viruses, either alone or in combination with carbohydrates or fats (glycoproteins and lipoproteins), figure prominently in the serologic and immunologic phenomena of viruses, such as neutralization. This is partly because proteins are good antigens (i.e., they elicit formation of large quantities of antibodies having strong affinity for the antigen) and because they comprise such a large part of the structure presented to cells in viral invasion.

The earliest studies on the chemistry of viruses were done on simple plant viruses, which proved, when adequately separated from cellular constituents, to be singularly lacking in readily identified enzymatic components. This is not to say that such viruses do not induce the formation of specific enzymes during their infectious cycles, but rather that such enzymes seem not to become components of the mature virus particles. The more complex viruses now appear to contain integral enzymatic components. These often fit into one of two classes of enzymes: (1) enzymes that attack cell walls or cell membranes, such as phage lysozyme or influenza neuraminidase; and (2) enzymes that catalyze the synthesis of nucleic acids, such as the RNA polymerases of reovirus or of the myxoviruses (influenza, Newcastle disease, etc.). Such enzymes are important protein components of the viruses which have them.

A function of viral structural proteins that doubtless confers a selective advantage for the survival of viruses is the protection they afford to the viral genome, i.e., the viral nucleic acid. Nucleic acids are subject to inactivation or degradation by shearing, by various chemicals and nuclease enzymes, and by radiations, especially by ultraviolet light. However, virtually all virus particles are constructed in such a way that the nucleic acid is inside a protein coat or is interwoven with protein at some distance from the particular surface. Hence the viral genome receives from its associated protein(s) considerable protection from physical and chemical degradative agents.

3-9 Primary Structure of Proteins

DETAILED INFORMATION ABOUT VIRAL PROTEINS IS OBTAINED BY DETERMINING THEIR AMINO ACID SEQUENCES

According to the central dogma relating the structure of nucleic acids to the structure of proteins, the polypeptide chain of a given protein has a specific sequence of amino acids which is dictated by the sequence of nucleotides in a particular gene. Some viral proteins have been sequenced in order to evaluate the mutational process and to relate structures to various viral functions.

There is no reliable method for determining the sequence of long polypeptide chains starting at one end and proceeding to the opposite end. However, it is possible to break such long chains into shorter ones by treatment with certain enzymes (trypsin is very useful) or chemicals (acid, cyanogen bromide, etc.). Such peptidic fragments can then be separated by a variety of techniques such as column chromatography, gel filtration, or electrophoresis. The small peptides are then sequenced by a variety of methods, outstanding among which is the stepwise degradation from the amino terminal and developed by Edman, which reached a high stage of performance with the production of an automatic "sequenator" (see Edman and Begg, 1967). Even under the best conditions, determining the sequence of amino acids in a protein is a major task; some of the details can be found in reviews such as the one by Fraenkel-Conrat and Rueckert (1967).

The proteins of a few strains of tobacco mosaic virus have been completely sequenced, and numerous others in part. In addition, the coat proteins of some small bacteriophages and the lysozymes of T4 phage have been sequenced. The first viral protein to be sequenced, and at that time (1960) the largest protein of any kind to be completely sequenced, was the coat protein of TMV. Total sequences, differing only slightly, were reported independently by a group in Schramm's laboratory in Tübingen, Germany, and a group in Stanley's laboratory in Berkeley, California. Subsequently the few differences were resolved as a consequence of refined analyses; the sequence now accepted in both laboratories is presented in Fig. 3-9 (see Hennig and Wittmann, 1972, for a review of analyses made by several laboratories on strains of TMV).

There are 158 amino acid residues in the TMV coat protein. Some viral proteins are larger than this, while the coat protein of the fl-fd-M13 group of long, flexuous, DNA-containing coliphages has only 49 amino acid residues. The fd protein, aside from its remarkably small size, is notable for its lack of arginine, all of its basic amino acid content being present as lysine asymmetrically distributed along the polypeptide chain. It also lacks cysteine and histidine, but these amino acids are often missing in viral proteins.

1 5 10 15
AcetylSer - Tyr - Ser - Ile - Thr - Thr - Pro - Ser - Gln - Phe - Val - Phe - Leu - Ser - Ser - Ala - Trp -
20 25 30 35
Ala - Asp - Pro - Ile - Glu - Leu - Ile - Asn - Leu - Cys - Thr - Asn - Ala - Leu - Gly - Asn - Gln - Phe -
40 45 50
Gln - Thr - Gln - Gln - Ala - Arg - Thr - Val - Val - Gln - Arg - Gln - Phe - Ser - Gln - Val - Trp - Lys -
55 60 65 70
Pro - Ser - Pro - Gln - Val - Thr - Val - Arg - Phe - Pro - Asp - Ser - Asp - Phe - Lys - Val - Tyr - Arg -
75 80 85
Tyr - Asn - Ala - Val - Leu - Asp - Pro - Leu - Val - Thr - Ala - Leu - Leu - Gly - Ala - Phe - Asp - Thr -
90 95 100 105
Arg - Asn - Arg - Ile - Ile - Glu - Val - Glu - Asn - Gln - Ala - Asn - Pro - Thr - Thr - Ala - Glu - Thr -
110 115 120 125
Leu - Asp - Ala - Thr - Arg - Arg - Val - Asp - Asp - Ala - Thr - Val - Ala - Ile - Arg - Ser - Ala - Ile -
130 135 140
Asn - Asn - Leu - Ile - Val - Glu - Leu - Ile - Arg - Gly - Thr - Gly - Ser - Tyr - Asn - Arg - Ser - Ser -
145 150 155 158
Phe - Glu - Ser - Ser - Ser - Gly - Leu - Val - Trp - Thr - Ser - Gly - Pro - Ala - Thr

FIG. 3-9. **Amino acid sequence of the coat protein of tobacco mosaic virus.**

3-10 *Viral Lipids*

PHOSPHOLIPIDS ARE MAJOR CONSTITUENTS OF VIRAL LIPIDS AND PLAY A STRUCTURAL ROLE IN MANY VIRUSES

A wide variety of fatty (lipid) compounds has been found in purified viruses, including phospholipids (phosphatides), glycolipids, neutral fats, fatty acids, fatty aldehydes, and cholesterol. However, phospholipid is the predominant type of fatty substance found in viruses that contain lipid, as indicated in Table 3-10.

Most of the lipid-containing viruses are characterized by an outer membranous coat, usually termed *envelope,* inside of which are located the viral genome and associated protein. The phospholipid of these viruses is located in the envelope and is important in the structural integrity of the virus particles. This is suggested by the pro-

TABLE 3-10 **Some Viruses That Contain Lipid**

Virus	*Lipid, %*	*Lipid constituents*
Avian myeloblastosis	35	Partly phospholipid
Equine encephalomyelitis	54	Phospholipid, cholesterol, neutral fat
Fowl plague	25	Phospholipid, cholesterol
Fowlpox	27	Phospholipid, cholesterol, neutral fat, fatty acids
Herpes simplex	22	Phospholipid
Influenza	18	Phospholipid, cholesterol, and miscellaneous fatty substances
Newcastle disease	27	Partly phospholipid
Potato yellow dwarf	20	Phospholipids and possibly other lipids
Pseudomonas phage PM2	10	Phospholipid
Rous sarcoma	35	Partly phospholipid
Sindbis	44	Phospholipid, cholesterol
Tipula iridescent	5	Phospholipid
Tomato spotted wilt	19	Not yet known
Vaccinia	5	Phospholipid, cholesterol, neutral fat

nounced loss of infectivity observed when such viruses are treated with fat solvents (ether, methanol-chloroform, deoxycholate, etc.) or with certain phospholipase enzymes. More directly, such viruses as the influenza virus have been disjoined into various component parts by treatment with ether (Hoyle's procedure). In practice, testing for infectivity after treatment with 20 percent ether has become a standard method for investigating the presence of an envelope structure. If infectivity is lost, it is considered strongly indicative (but not proof) that the virus has an envelope, because most viruses that lack a lipid component are insensitive to ether treatment. All the viruses listed in Table 3-10, except for *Tipula* iridescent virus, are sensitive to fat solvents; the sensitive viruses contain what Franklin has termed "peripheral structural lipids," whereas the lipid of *Tipula* iridescent virus either is not structural or is internally located.

It is not known precisely why removal of lipid destroys the infectivity of some viruses. Among the possibilities are the following:

1 Lipid or lipoprotein may be required for attachment of virus to cells as the initial step in infection.
2 The nucleoprotein in the interior of enveloped viruses may be exceptionally susceptible to destruction by enzymes if exposed at the wrong time—in this case, at the outset of the infectious process.
3 The fat solvent may do more than extract lipid; it may also denature protein, let us say of the nucleoprotein inside the envelope, thus blocking release of the nucleic acid and performance of its genetic functions.

VIRAL LIPIDS PROBABLY ORIGINATE FROM HOST CELL MEMBRANES

Some years ago the problem of origin of viral phospholipids was attacked by Schäfer and associates, using fowl plague virus as a model for enveloped viruses. Cultures of embryonic chicken cells were given identical amounts of ^{32}P but at different times (0 to 48 hr) before infection with the virus. The virus was subsequently isolated, and the lipid and RNA fractions were analyzed for ^{32}P content. It was found that the ^{32}P content of the viral lipid fraction increases with the length of time prior to infection at which the isotope is added, whereas the ^{32}P content of the viral RNA is virtually independent of the time at which the ^{32}P is supplied to the cells (Fig. 3-10). Hence it appears that the phospholipid of the virus is synthesized before infec-

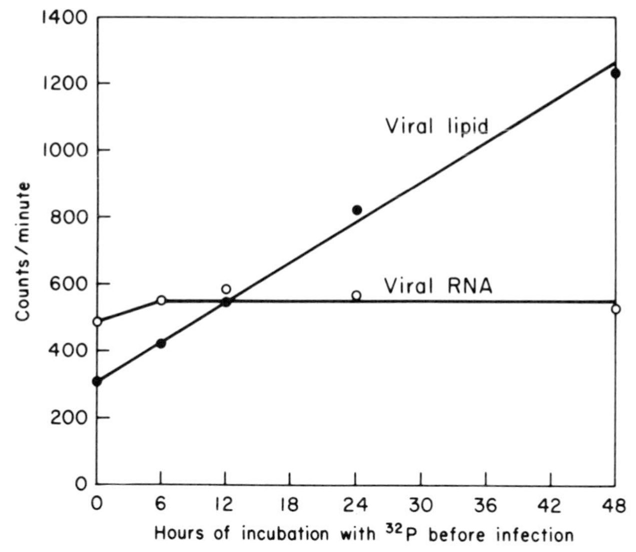

FIG. 3-10. **Incorporation of ^{32}P into the phospholipid and RNA of fowl plague virus as a function of time of exposure of embryonic chicken cells to the isotope prior to infection with virus. The same quantity of isotope was administered at each time interval, and the resultant incorporation is expressed in terms of counts per minute for a fixed amount of virus. (From Schäfer, Perspect. Virol. 1, 34, 1959.)**

tion, whereas the viral RNA is of course made only after infection. This conclusion needs a slight revision in view of the more detailed analyses made by Kates and associates on phosphorus distribution in influenza virus–infected cells. The results of these analyses indicate that the lipids of influenza virus are mostly derived from lipids present in the host cell before infection but that a small portion of viral phospholipid comes from material newly synthesized (but not virus-directed) by the infected cells.

Examination in the electron microscope of ultrathin sections of chick embryo cells infected with fowl plague virus shows no complete virus particles in the interior of the cells. However, spherical particles of the proper size for fowl plague virus are observed in cell wall protrusions. The particle-containing protrusions are most evident in cell sections at times after infection when infectivity extractable from such cells or found in the surrounding culture medium is rising.

Such facts lead to the conclusion that fowl plague virus particles are completed at the cell membrane and acquire some host lipid as they are extruded from the cell. This view is generally held for all enveloped viruses, although some enveloped viruses, such as potato yellow dwarf virus, may acquire lipid from nuclear rather than from cytoplasmic membrane. In any case, it should not be supposed that viral lipids occur as such in the virus particle. There is evidence that the lipid is combined with carbohydrate in some instances to form glycolipid, and there may be some combined with proteins to yield lipoprotein. The precise nature of the host contribution in enveloped viruses remains to be clarified.

3-11 Viral Carbohydrates

ALL VIRAL NUCLEIC ACIDS HAVE CARBOHYDRATE MOIETIES, BUT IN ADDITION SOME VIRUSES HAVE GLUCOSYLATED PYRIMIDINES AND SOME OTHERS HAVE GLYCOPROTEINS

All viruses contain carbohydrate, since nucleic acid is an indispensable constituent of viruses and nucleic acid is composed in part of either ribose or deoxyribose.

In addition to nucleic acid carbohydrate, some bacterial viruses (such as the T2, T4, and T6 phages of *E. coli*) have glucose attached by an 0-glycosidic link to the 5-hydroxymethylcytosine (5-HMC) of the viral DNA.

Although the T-even phages contain essentially the same quantity of 5-HMC in their DNA components, the amount of glucose attached is distinct for each, amounting to 0.8, 1.0, and 1.6 moles of glucose per mole of 5-HMC for T2, T4, and T6, respectively. Furthermore, the distribution of glucosyl residues and the configuration of the linkage between glucose and 5-HMC (whether α or β) vary among these phages, as summarized in Table 3-11.

As indicated in the table, about one-fourth of the 5-hydroxymethylcytosine residues of T2 and T6 DNAs are not glucosylated, whereas all those of T4 are. The diglucosyl groups, found in a high proportion in T6 DNA, represent a disaccharide which has been identified as gentiobiose.

There is evidence from genetic recombination experiments involving crosses between T-even phages that the degree of glucosylation is an inherited trait, although the trait "glucose content" does not

TABLE 3-11 **Distribution of Glucosyl Residues Attached to 5-Hydroxymethylcytosine in the DNA of T2, T4, and T6 Bacteriophages**

Phage	*Unglucosylated,%*	*Monoglucosyl α linkage, %*	*Monoglucosyl β linkage, %*	*Diglucosyl α,β linkage, %*
T2	25	70		5
T4	0	70	30	0
T6	25	3		72

SOURCE: Compiled from data of Lehman and Pratt, *J. Biol. Chem.* 235, 3254–3259, 1960.

segregate symmetrically as a simple Mendelian character. For example, the recombinants of T2 × T6 cross were found to have the glucose content of either T2 or T6, but many more were found to have the glucose content of T2 and the host range of T6 than vice versa.

The precise function of the glucosyl residues is in any case unknown. However, certain enzymes are required to glucosylate 5-HMC during synthesis of viral nucleic acid, and once it is synthesized, the presence of glucosyl units appears to confer resistance to degradation of such DNA by some diesterases.

It should be noted that at least two *Bacillus subtilis* phages, PBS1 and SP8, have also been found to contain glucosylated DNA, even though they do not contain 5-HMC. The glucosyl units appear to be attached to the cytosine and guanine rings in these cases, although the details of this attachment have not been well worked out.

By means of the acrylamide gel technique coupled with the use of radioactive sugars, it has been shown that viruses with an envelope structure often contain some of their proteins in the form of glycoproteins. These glycoproteins consist of proteins to whose serine, threonine, and asparagine residues carbohydrates are glycosidically linked. The carbohydrate components are complex polysaccharides usually fabricated from fucose, galactose, glucosamine, and mannose.

The enveloped paramyxoviruses, simian virus 5, Newcastle disease virus, and Sendai virus each contain two glycoproteins. These appear to occur in the morphologic projections called *spikes,* to which hemagglutinating and neuraminidase activities are also ascribed. A similar situation exists with influenza and other myxoviruses. However, the function of spikes on such viruses as sowthistle yellow vein virus is not yet clear, and many other enveloped viruses (such as herpes-, Sindbis, ϕ6, and tomato spotted wilt viruses) have no spikes.

The surface location and antigenic nature of glycoproteins in the particles of enveloped viruses make them serologically active, and there is evidence that they are important in determining the specificity among viruses of a given type, such as the influenza viruses. Whether any of the glycoproteins are derived from host cell membranes has not been determined, although it is suspected that some may be.

3-12 Miscellaneous Viral Constituents

POLYAMINES AND METALS ARE FOUND IN SOME VIRUSES

Polyamines such as putrescine and spermidine are widely distributed in cells of various kinds, and they have also been detected as components of some viruses. At common cellular pH values the polyamines are protonated; hence they are polycations. Such cations have an affinity for the thousands of phosphoryl anions of nucleic acids and hence can and do combine with nucleic acids as a consequence of electrostatic attractions.

Some plant viruses, such as turnip yellow, turnip crinkle, and broad-bean mottle viruses, contain small amounts of polyamine, as do certain *E. coli* phages. There seems to be a complete lack of specificity with respect to a particular polyamine in the case of the phages. For example, Ames and associates found that putrescine and spermidine were present in both phage T4 and its host, *Escherichia coli.* However, if *E. coli* was grown in a medium rich in spermine, a polyamine not usually found in *E. coli,* the putrescine and spermidine normally found in T4 were replaced by spermine and acetylated spermine without any demonstrable change in phage function. Furthermore, it was observed that a permeable mutant of T4 could be washed with a magnesium salt, displacing essentially all the polyamine with Mg ions without any change in infectivity.

Metallic cations have been found in most viruses where they have been sought. For example, Loring and associates found iron, calcium, magnesium, copper, and aluminum rather tightly bound to tobacco mosaic virus. These cations seem mainly associated with the nucleic acid of the virus, from which they can be largely dissociated by treatment with a chelating agent without significantly reducing infectivity.

The general conclusion about both organic and inorganic cations is that they bind randomly to viral nucleic acids in amounts depending on environment and relative affinities of the ions involved. This binding may well affect the conformation of the viral nucleic acid, but since there seems to be no essential specificity in the nature of the cations bound, they are generally considered to be variable, adventitious constituents of viruses.

3-13 *Morphology*

VIRUS PARTICLES ARE SPHEROIDAL, ELONGATED, OR COMBINATIONS OF THESE SHAPES

The chemical constituents that were described in the previous section are in nature fabricated into virus particles of diverse sizes and shapes depending upon the virus in question. There are several ways to classify the resulting particles; the simplest distinction is based on the images of viruses revealed by the electron microscope: (1) spheroidal particles (also called spherical or isometric particles), (2) elongated particles, (3) combination particles (such as a phage particle that has a spheroidal head and an elongated tail). Some viruses, especially enveloped viruses, tend to be pleomorphic, in which case both spheroidal and elongated particles may appear in a given population of particles.

It is possible to obtain evidence regarding the size and shape of virus particles by indirect methods. For example, solutions or suspensions of viruses can be subjected to measurements of flow birefringence, viscosity, filterability through membranes of known porosity, diffusion rate, sedimentation rate, and light-scattering capacity, including the scattering of x-rays. The main contribution of each of these methods to the characterization of viruses is indicated in Table 3-12. Historically, these were the tools used to characterize viruses before the electron microscope became available as a research instrument. Many of these techniques are still used to good advantage in studying viruses, but the most versatile and direct method for determining the architecture of virus particles is examination in the electron microscope.

Numerous ingenious techniques have been devised for electron microscopy that increase both the accuracy and amount of information provided. Some of these techniques enhance contrast between virus particles and the plastic film of the microscope mount, some minimize the tendency of particles to collapse when exposed to osmotic and surface tension forces, and others limit the destructive

TABLE 3-12 **Indirect Methods for Evaluating Size and Shape of Viruses**

Method	*Distinguishes primarily*
1. Flow birefringence	Shape
2. Viscosity	Shape
3. Ultrafiltration and gel chromatography	Size
4. Diffusion	Size
5. Centrifugation:	
Analytical	Size, shape, and density
Rate-zonal	Size, shape, and density
Isopycnic gradient	Density
6. Light scattering:	
Visible light	Size
X-ray	Size, configuration

effects of the beam of electrons used to illuminate the field under examination. No attempt will be made here to describe these methods, which can be found in books such as those by Kay, 1961, and Huxley and Klug, 1971, but rather some of the results of their application will be used to illustrate the structure of viruses.

First, something must be said about terminology. As more and more information was obtained about the chemistry and structure of virus particles it was necessary to employ terms to describe the structures observed. This led to a variety of terms and eventually precipitated attempts to standardize the nomenclature. Such terms as the following have issued: *virion,* the mature, potentially infectious virus particle; *capsid,* the protein coat built around and closely associated with the viral nucleic acid, the two together being called *nucleocapsid; structure units,* the protein subunits of the capsid; *capsomers* or *morphologic units,* capsid substructures distinguishable in the electron microscope and representing either individual structure units or clusters of them; *envelope,* the structure consisting of lipid, protein, and carbohydrate which usually envelops the nucleocapsid of viruses that mature at cell membranes. Some of these terms seem needlessly recondite, but since they are in the viral literature, Table 3-13 is presented in order to indicate synonymous expressions. In this table, the older, simpler terms are listed in the left column. Even these simple terms need considerable qualification when applied to the structurally more complex viruses, as will be seen in the following discussion.

TABLE 3-13 **Viral Terminology**

Term	*Synonyms*
1. Virus	Virion; virus particle
2. Protein coat	Capsid; protein shell
3. Nucleoprotein	Nucleocapsid; NP
4. Protein subunit	Structure unit
5. Morphologic unit	Capsomer
6. Envelope	Peplos
7. Spike	Peplomer

SPHEROIDAL VIRUSES. HIGH-RESOLUTION ELECTRON MICROSCOPY SHOWS THAT THE PARTICLES OF SPHEROIDAL VIRUSES ARE SYMMETRIC POLYHEDRA

As mentioned earlier, a general principle of virus architecture is that much of the mass of the particle consists of a protein shell and this protein is fabricated from many identical subunits. Caspar and Klug have stressed that a virus particle so constructed will have a uniform size and regular shape. Furthermore, x-ray diffraction and electron microscope data combined with geometry and model building indicate that most spheroidal viruses are organized according to a type of cubic symmetry called *icosahedral symmetry.* According to the mathematic rules of icosahedral symmetry, modified somewhat by the theory of quasiequivalence developed by Caspar and Klug, it can be predicted that spheroidal viruses will have specific numbers of morphologic units. Some examples of the classes according to number of morphologic units, and some viruses possibly illustrating the classes, are given in Table 3-14. For comparison, the numbers of protein subunits (structure units) are also given in the table as a reminder that the units visualized in the electron microscope usually consist of more than one of the fundamental protein subunits. In the Caspar and Klug concept, the protein subunits of an icosahedral virus may be thought to occur in groups of five (pentamers) and six (hexamers), as indicated by the examples in Table 3-14 (however, see footnote to table). In order to have them identically situated, there can be no more than 60 identical subunits on the surface of a sphere, but with quasiequivalence there may be larger numbers, which, however, will be multiples of 60, as indicated in the table.

TABLE 3-14 **Possible Numbers of Morphologic Units and Subunits in Virus Particles Having Icosahedral Symmetry***

No. of morphologic units	*No. of subunits*	*Grouping of subunits in forming morphologic units*†	*Virus example*
12	60	12 pentamers	Coliphage φX174
32	180	12 pentamers 20 hexamers	Broad-bean mottle, cowpea chlorotic mottle, cucumber mosaic, turnip yellow mosaic
42	240	12 pentamers 30 hexamers	Arabis mosaic, tobacco ring spot
72	420	12 pentamers 60 hexamers	Human wart, polyoma, simian virus 40, Shope papilloma
90	180	90 dimers	Tomato bushy stunt, turnip crinkle
92	540	12 pentamers 80 hexamers	Reovirus, wound tumor
162	960	12 pentamers 150 hexamers	Herpes simplex, varicella
252	1,500	12 pentamers 240 hexamers	Adenovirus, infectious canine hepatitis

* There are classes of icosahedral particles other than those listed here, but they were omitted for lack of virus examples to illustrate them. See Caspar and Klug (1962) for a detailed discussion.

† These groupings of subunits are conceptual and may or may not coincide with the actual situation. For example, coliphage φX174 seems to have four different protein components rather than 60 copies of one, and the precise numbers and morphologic arrangement of the four proteins remain to be worked out. Similarly, adenovirus has several different protein components, of which the major coat constituent, the hexon, probably consists of three polypeptides, which, moreover, are not identical.

In all this, an essential point to remember, as Caspar and Klug pointed out, is that the structures assumed by virus particles, spheroidal or otherwise, reflect the requirement that these structures assume a minimum energy configuration.

1. Some Small Spheroidal Viruses. One of the difficulties encountered in the electron microscopic examination of virus particles is that there is often little difference in the electron scattering of (and hence contrast between) the virus particles and the plastic film supporting them. Much of this difficulty is overcome by the technique of Williams and Wyckoff, in which an electron-dense material such as gold, palladium, or molybdenum is vaporized *in vacuo* at an oblique angle against the virus particles (this is called *shadow casting*). The shadowing material coats the virus particles but leaves more lightly coated regions (shadows) on the film next to the virus particle on the side farther from the source of shadowing material. Thus the contrast between particle and film is enhanced. However, the fine details of surface structure are lost by this coating procedure. Fortunately, another technique, called *negative staining* (see Horne, 1963), has been developed which nicely complements the shadow-casting method. In negative staining, the virus is suspended in an electron-dense material (sodium or potassium phosphotungstate) and then is either deposited in a small droplet or sprayed on the specimen mount. As the mount dries, the phosphotungstate drains down from around the virus particles and surrounds them in such a manner as to reflect surface structures, hollow regions, etc. The mounts are then examined in the electron microscope. Most of the micrographs shown here to illustrate the structure of different viruses were made by the negative staining technique.

Some small, spheroidal plant, animal, and bacterial viruses are illustrated in Fig. 3-11. Morphologic units are apparent in all of them but show more clearly in some than in others. The apical morphologic units of phage ϕX174 (Fig. 3-11*g*) are clearly discernible. One or more of these, it has been postulated, may serve in attachment of the virus to susceptible bacterial cells. However, no such outstanding structures are apparent on the particles of a similar phage, Qβ (Fig. 3-11*h*). Dark centers can be observed in some of the particles shown (Fig. 3-11*a*, *b*, *h*); these may represent particles which lack a full complement of nucleic acid and hence allow phosphotungstate to drain through them as well as around them and leave a heavier deposit under the particles.

2. Some Large Nonenveloped Spheroidal Viruses. A variety of larger spheroidal viruses is illustrated in the electron micrographs of Fig. 3-12. The morphologic units of several of these viruses are quite distinct, and some curious additional features can be noted with some of them. Adenovirus (Fig. 3-12*b*) has some bizarre structures called *penton fibers* which project from certain points of the particle. Reovirus is notable in having two protein coats, an outer one, and

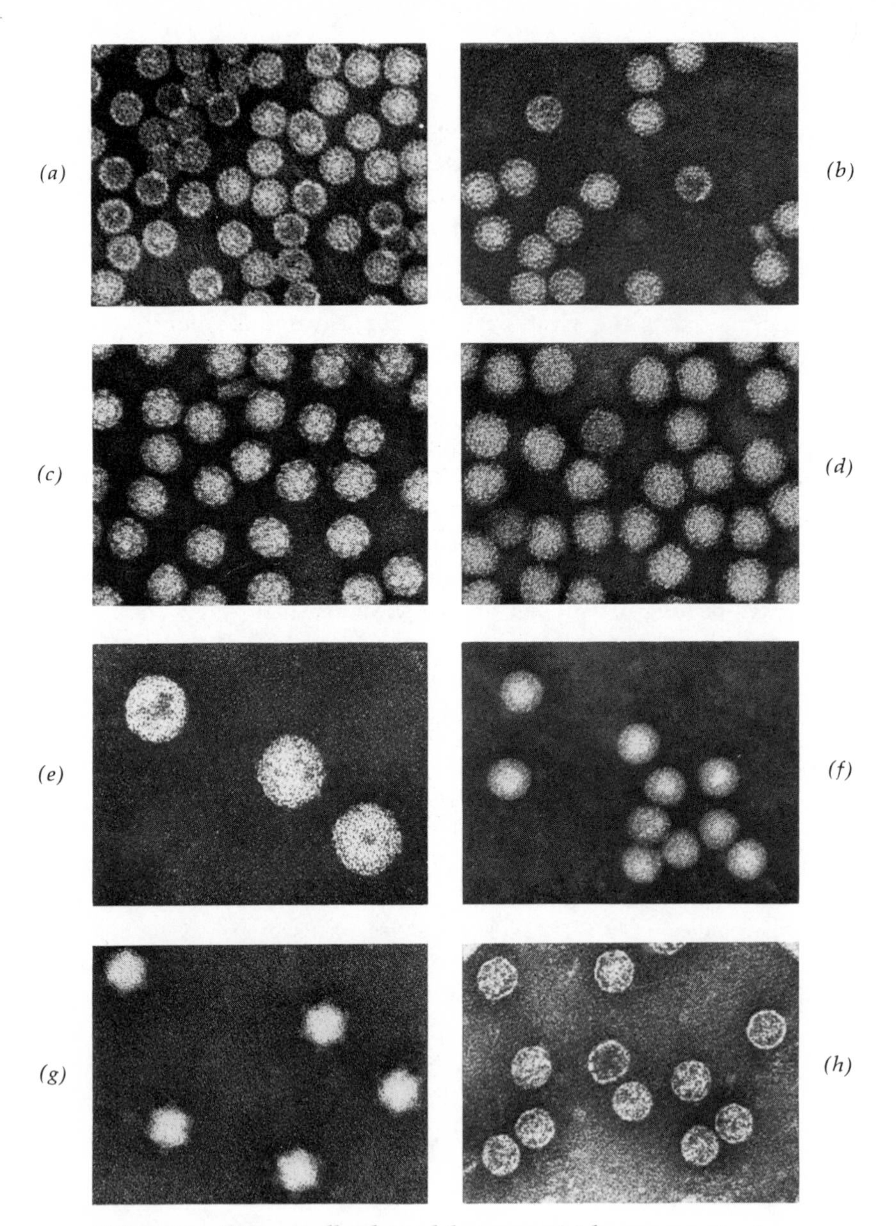

FIG. 3-11. **Some small spheroidal viruses: (a) brome mosaic virus; (b) cowpea chlorotic mottle virus; (c) turnip yellow mosaic virus; (d) tomato bushy stunt virus; (e) cauliflower mosaic virus; (f) poliovirus; (g) ϕX174 coliphage; (h) Qβ coliphage. Mounts were prepared by the negative staining technique. ($\times$ about 140,000.) Most of these viruses are about 30 nm in diameter except for cauliflower mosaic virus, which is about 50 nm. (Courtesy of R. C. Williams and H. W. Fisher.)**

an inner one which is often referred to as the core. A core can be seen in the top particle of Fig. 3-12*d*. A suggestion of internal structure is evident in the micrographs of the cytoplasmic polyhedrosis virus (Fig. 3-12*e*).

3. Some Large Enveloped Spheroidal or Elongated Viruses. Viruses whose structures are completed at cell membranes usually have envelope structures, parts of which are contributed by the host membrane. Such structures generally display protuberances, called *spikes* (or *peplomers*), which can often be visualized in the electron microscope when the negative staining technique is used. Enveloped viruses, perhaps because of their lipid content, seem to be more pliable than other viruses and hence may show pleomorphic character. Some elongated forms of influenza virus are shown in Fig. 3-13, along with the common spheroidal particles. Some other examples of enveloped viruses are given in Fig. 3-13, including an avian RNA tumor virus and the plant virus sowthistle yellow vein virus. The latter is characterized by large, bacilliform particles which seem to possess membranous structures. There is a remarkable similarity between the structure of the sowthistle virus particles and those of vesicular stomatitis virus in the animal virus series. Sometimes bullet-shaped particles are seen, but these may well be artifacts derived by partial rupture of the bacilliform particles. Though this is rather certain for the sowthistle virus and may be so for about 15 other plant viruses of this sort, it does not seem to be the case for the comparable vertebrate viruses, for which the bullet-shaped particle appears to be truly characteristic.

4. Large Brick-Shaped Viruses. Poxviruses are the largest and most complex of the viruses. The particles of these viruses are usually described as brick- or loaf-shaped. When they are treated with phosphotungstate and viewed in the electron microscope, a whorled pattern of tubules or filaments is discernible, as shown in Fig. 3-14*a*. A combination of enzymatic and chemical treatments with electron microscopy has shown other structures lying beneath the filamentous surface. These include "lateral bodies," of unknown composition and function, and a core or nucleoid, consisting of the viral nucleic acid and a protein coat (Fig. 3-14*b* and *c*).

5. Elongated Viruses. Some elongated viruses, such as tobacco mosaic virus, have rodlike particles, while other viruses are characterized by flexuous, fibrillar particles. Two examples of each type are given

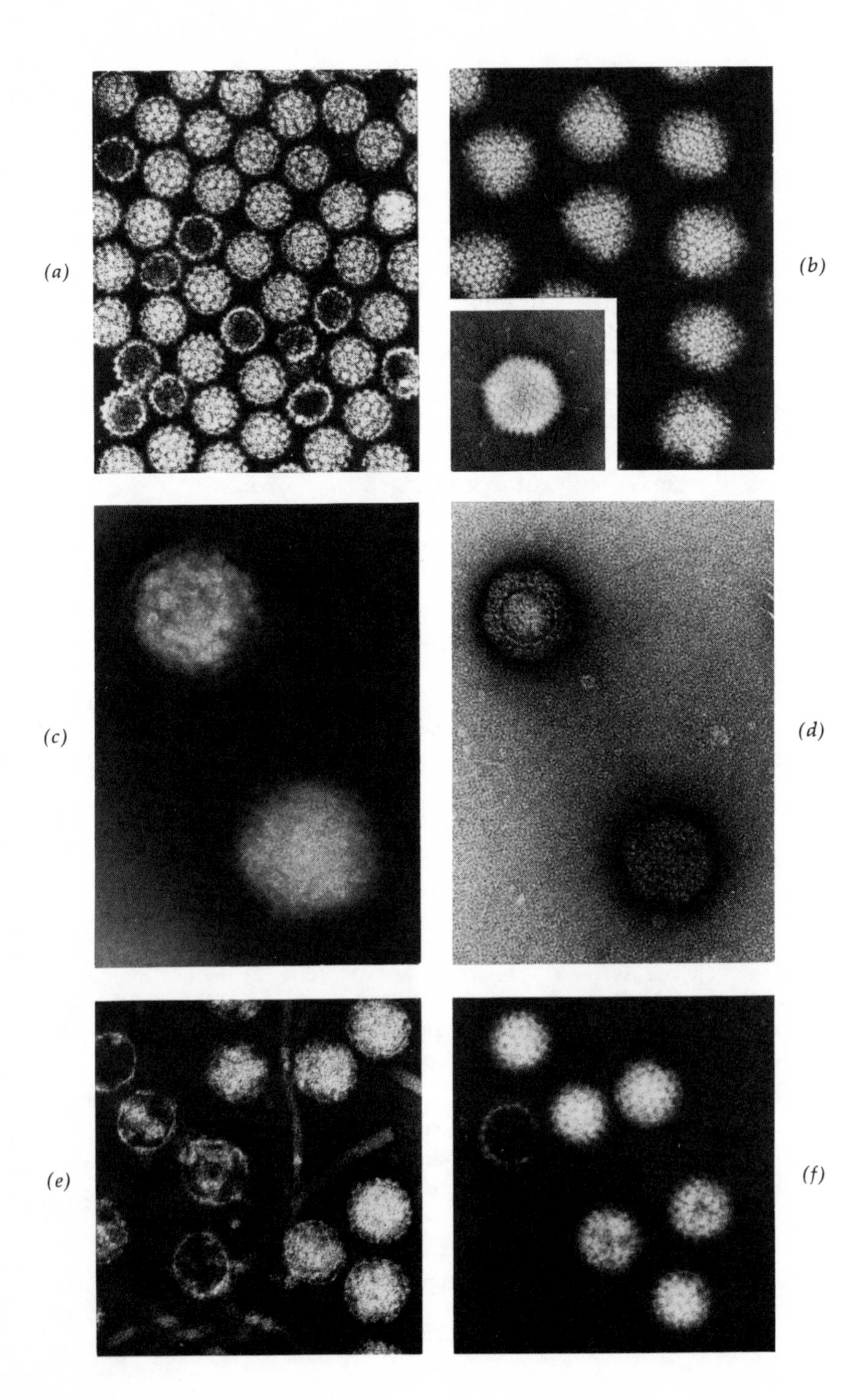
(a)
(b)
(c)
(d)
(e)
(f)

in Fig. 3-15. It should be noted that preparations of TMV contain particles of various lengths. This is common for TMV, but only the particles that are about 300 nm in length are infectious. Furthermore, in the infectious process the long particles do not require help from the short ones. In contrast, the short particles of tobacco rattle virus (Fig. 3-15*b*) are of uniform rather than variable length, and they are needed for infectivity. Part of the genome of tobacco rattle virus is in the long particles, and part in the short particles. This is discussed in more detail in Chap. 5. The negative staining reveals that the particles of both these viruses have a central hole running the length of the particles. The protein subunits can be discerned in the micrograph of TMV as striations perpendicular to the long axis of the particle. The particles of potato virus X and of fd phage are so long that complete particles cannot be shown at the magnification of Fig. 3-15.

6. Bacterial Viruses with Spheroidal Heads and Elongated Tails of Various Sizes and Shapes. The bacterial viruses probably exhibit the richest assortment of sizes and shapes of any class of viruses. Two small spheroidal phages are illustrated in Fig. 3-11, and an elongated one in Fig. 3-15. There are numerous other phages with spheroidal heads and a variety of elongated tails. Some of these are illustrated in Fig. 3-16.

7. Encapsulated Insect Viruses. Some insect viruses occur as particles that are similar to those of other animal viruses. One of these is the *Tipula* iridescent virus shown in Fig. 3-12. However, the majority of insect viruses appear to occur in their mature form in characteristic inclusion bodies. Inclusion bodies are crystalline packages which contain one or more virus particles. There are two major types of inclusion bodies: polyhedra and capsules (also called *granules*). The diseases with which polyhedral bodies are associated are called *polyhedroses,* and those involving capsules, *granuloses.* A further division of the polyhedroses is made on the basis of the site of virus multiplication; thus there are nuclear and cytoplasmic polyhedroses. The virus particles of nuclear polyhedroses are generally rod-shaped;

FIG. 3-12. **Some large nonenveloped spheroidal viruses: (a) Shope rabbit papilloma virus; (b) adenovirus; (c) Tipula iridescent virus; (d) reovirus; (e) cytoplasmic polyhedrosis virus of the silkworm; (f) wound tumor virus of sweet clover. Mounts were prepared by the negative staining technique. (× about 140,000.) These viruses range from about 50 to 130 nm in diameter. (Courtesy of R. C. Williams and H. W. Fisher.)**

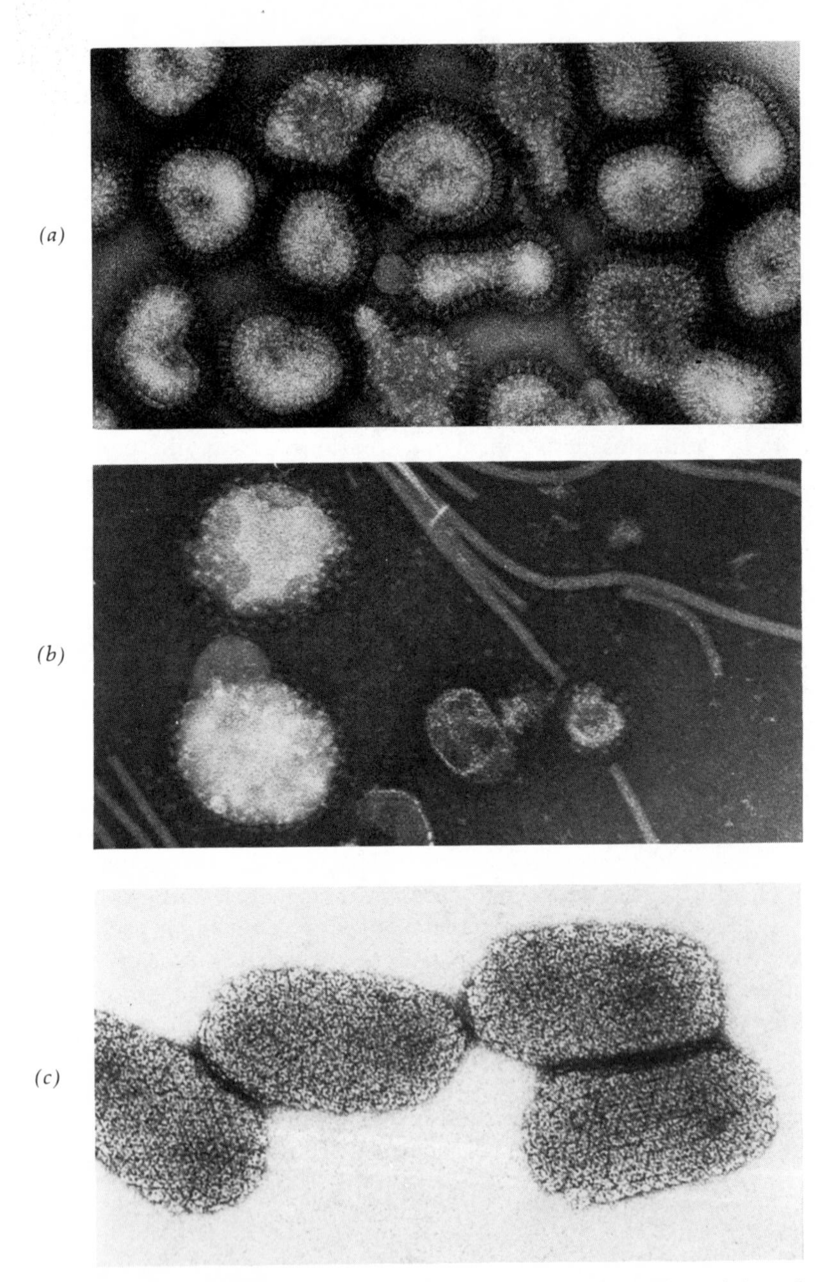

FIG. 3-13. **Some large enveloped viruses: (a) spheroidal and elongated particles of influenza virus (an example of pleomorphism); (b) Rous sarcoma virus; (c) sowthistle yellow vein virus. (The first two micrographs were kindly provided by R. C. Williams, and that of the sowthistle yellow vein virus by D. Peters.)**

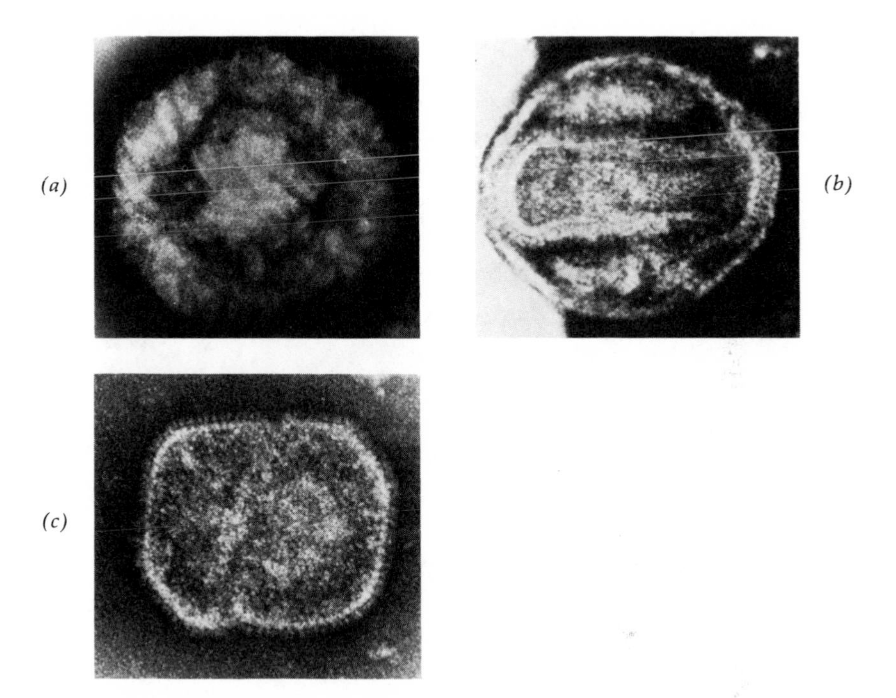

FIG. 3-14. **Vaccinia virus, a poxvirus. Mounts were prepared by negative staining. (a) Whole virus particle showing surface tubules; (b) a particle partly stripped with detergent so as to reveal core and lateral bodies; (c) a core showing regular surface structures. (a, courtesy of R. C. Williams and H. W. Fisher; b and c, courtesy of K. B. Easterbrook.)**

those of the cytoplasmic polyhedroses are spheroidal. Hundreds of virus particles are occluded in a crystalline protein matrix in polyhedral bodies. The virus particles can be released from crystalline polyhedral bodies by treatment with dilute alkali. Some particles of a cytoplasmic polyhedrosis virus of the silkworm, released by alkaline treatment of the polyhedral bodies, are shown in Fig. 3-12. Thin sections can be made of polyhedral bodies which upon electron microscopy reveal the dispersion of virus particles. Such sections of polyhedral bodies from infected cabbage loopers (*Trichoplusia ni*) are illustrated in Fig. 3-17.

In granuloses, a single rod-shaped particle of virus (occasionally two) is found in each capsule of crystalline protein. This is illustrated by thin sections of capsules from diseased meal moths (*Plodia interpunctella*) (Fig. 3-18). The single virus particle, which is in the shape

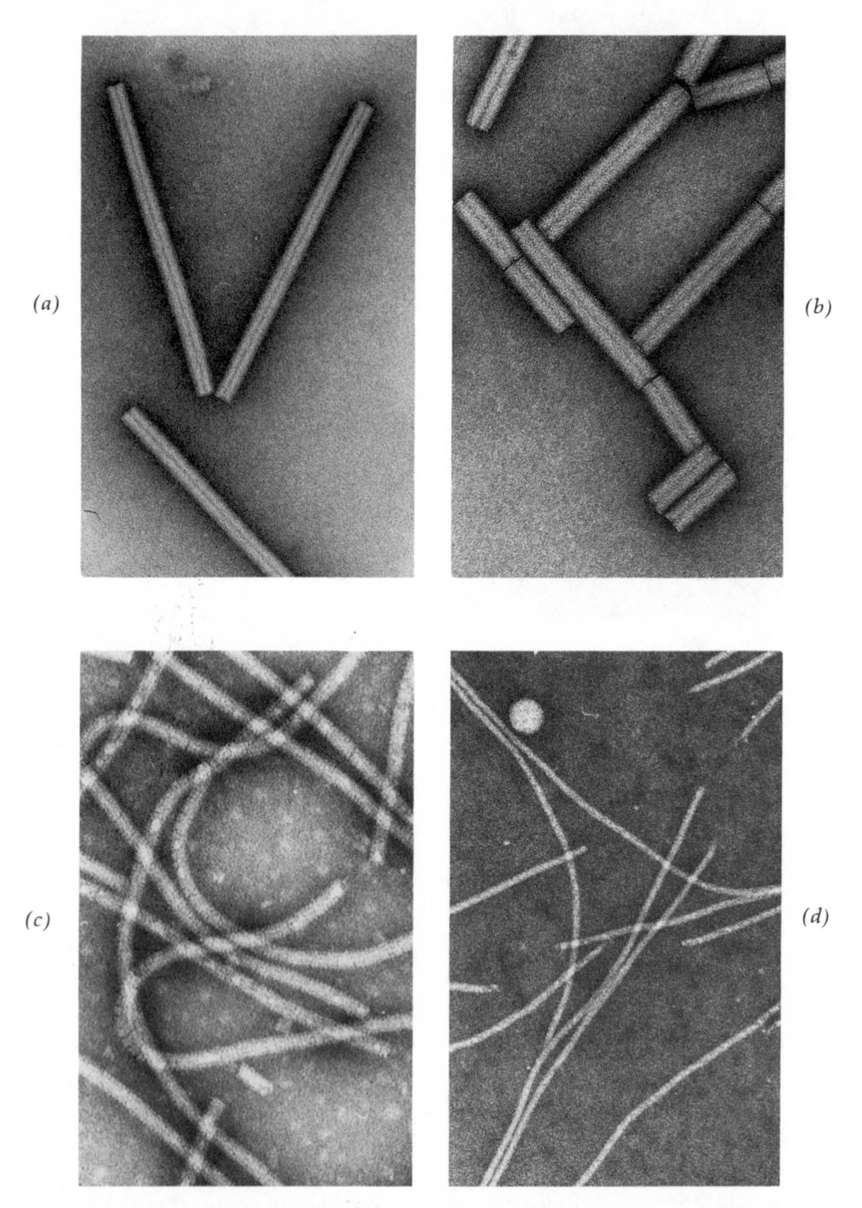

FIG. 3-15. **Some elongated viruses: (a) tobacco mosaic virus; (b) tobacco rattle virus; (c) potato virus X (latent mosaic of potato); and (d) coliphage fd. (Courtesy of R. C. Williams and H. W. Fisher.)**

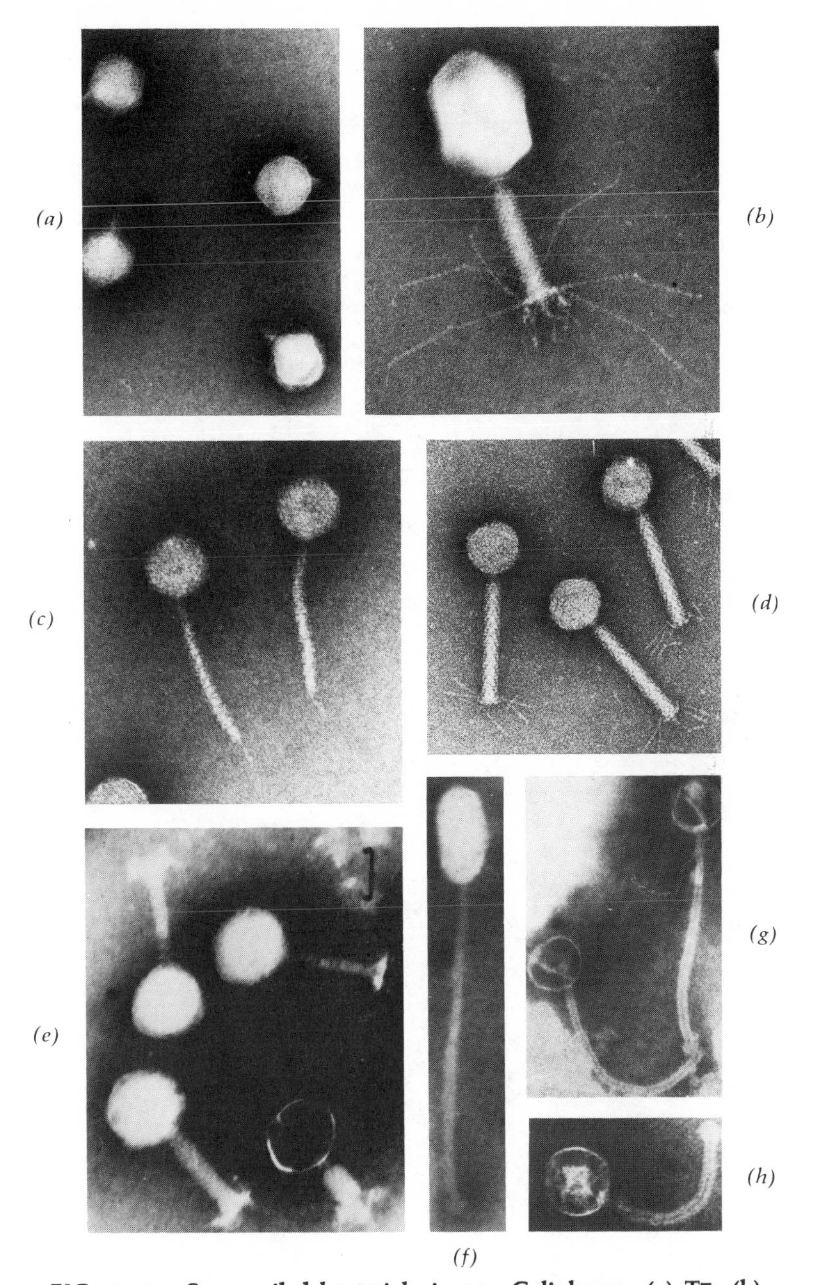

FIG. 3-16. **Some tailed bacterial viruses: Coliphages: (a) T7; (b) T4; (c) lambda; (d) P2; and (e) typhoid phage Vi 1; (f) Staphylococcus phage 6; (g) Staphylococcus phage 77; (h) Pseudomonas phage Pc; and (i) a Brucella phage. (Coliphages, courtesy of R. C. Williams and H. W. Fisher, and the rest, courtesy of D. E. Bradley and D. Kay.)**

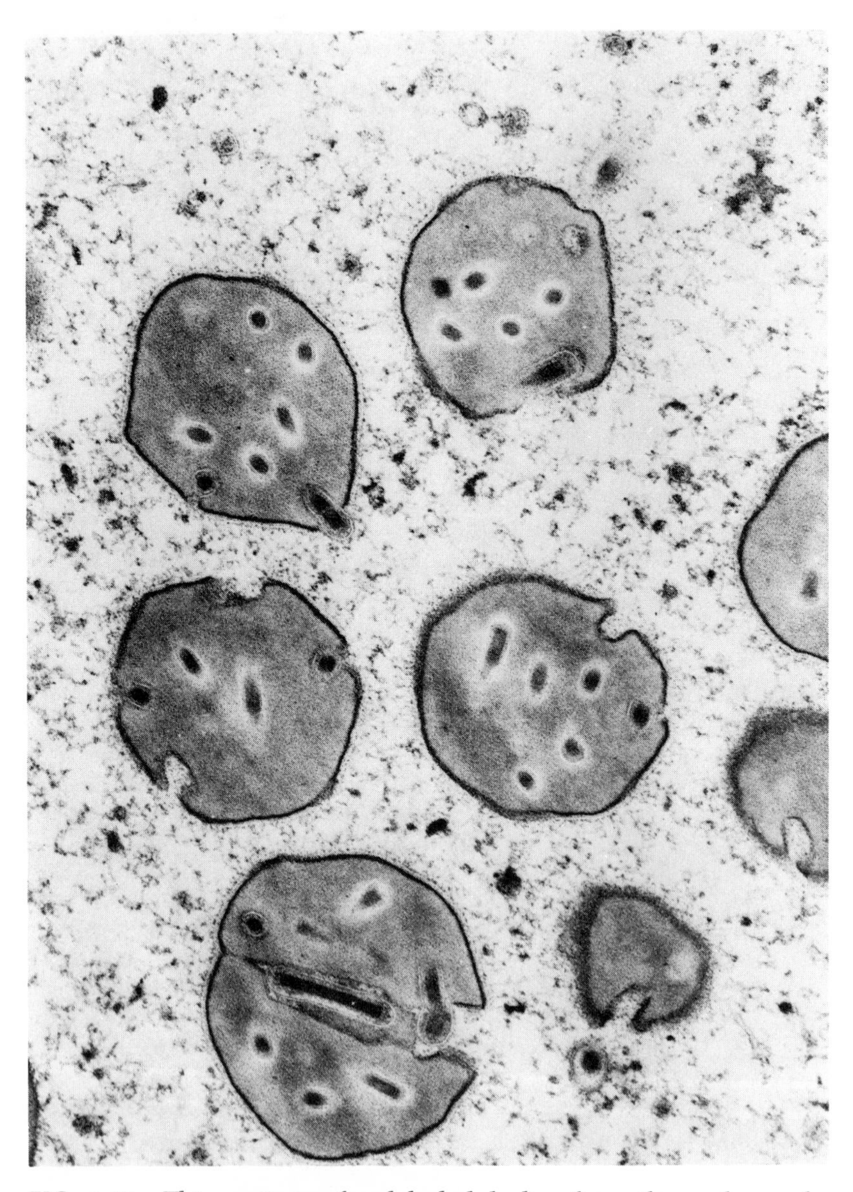

FIG. 3-17. **Thin sections of polyhedral bodies from the nuclear polyhedrosis of the cabbage looper (Trichoplusia ni). Since the viral rods are randomly oriented, one seldom sees a full-length particle in a thin section. One such particle is present in the sections of polyhedral bodies shown here; numerous parts of other virus particles are apparent. (Courtesy of M. D. Summers.)**

of a slightly curved rod (Fig. 3-18*a*), is centrally located in the crystalline capsule. Two concentric membranes surround the virus particle; the one closer to the virus particle is termed *intimate membrane,* and the other one is called *outer membrane.* These and the crystalline lattice surrounding them are nicely shown in Fig. 3-18*b*.

References

BOOKS

Fenner, F.: *The Biology of Animal Viruses,* vol. 1, chap. 3, Academic, New York, 1968.

Fraenkel-Conrat, H.: *The Chemistry and Biology of Viruses,* chaps. 5 to 7, Academic, New York, 1969.

——— (ed.): *Molecular Basis of Virology,* Reinhold, New York, 1968.

Huxley, H. E., and A. Klug: *New Developments in Electron Microscopy,* The Royal Society, London, 1971.

Kay, D.: *Techniques for Electron Microscopy,* Charles C Thomas, Springfield, Ill., 1961.

Knight, C. A.: *Chemistry of Viruses,* Springer-Verlag, Vienna, 1963.

Maramorosch, K., and E. Kurstak (eds.): *Comparative Virology,* Academic, New York, 1971.

Mathews, C. K.: *Bacteriophage Biochemistry,* Van Nostrand Reinhold, New York, 1971.

Matthews, R. E. F.: *Plant Virology,* Academic, New York, 1970.

Smith, K. M.: *Insect Virology,* Academic, New York, 1967.

Stent, G. S.: *Molecular Genetics,* chaps. 12 and 13, Freeman, San Francisco, 1971.

Tikhonenko, A. S. (translated from the Russian by Basil Haigh): *Ultrastructure of Bacterial Viruses,* Plenum, New York, 1972.

Watson, J. D.: *The Double Helix,* Atheneum, New York, 1968.

JOURNAL ARTICLES AND REVIEW PAPERS

Composition of Viruses

Fraenkel-Conrat, H., and R. R. Rueckert: Analysis of Protein Constituents of Viruses, in *Methods in Virology,* K. Maramorosch and H. Koprowski (eds.), vol. 3, pp. 1–75, Academic, New York, 1967.

Green, M.: Chemical Composition of Animal Viruses, in *The Biochemistry of Viruses,* H. B. Levy (ed.), pp. 1–54, Marcel Dekker, Inc., New York, 1969.

Harrison, B. D., J. T. Finch, A. J. Gibbs, M. Hollings, R. J. Shepherd, V. Valenta, and C. Wetter: Sixteen Groups of Plant Viruses, *Virology* **45**: 356–363, 1971.

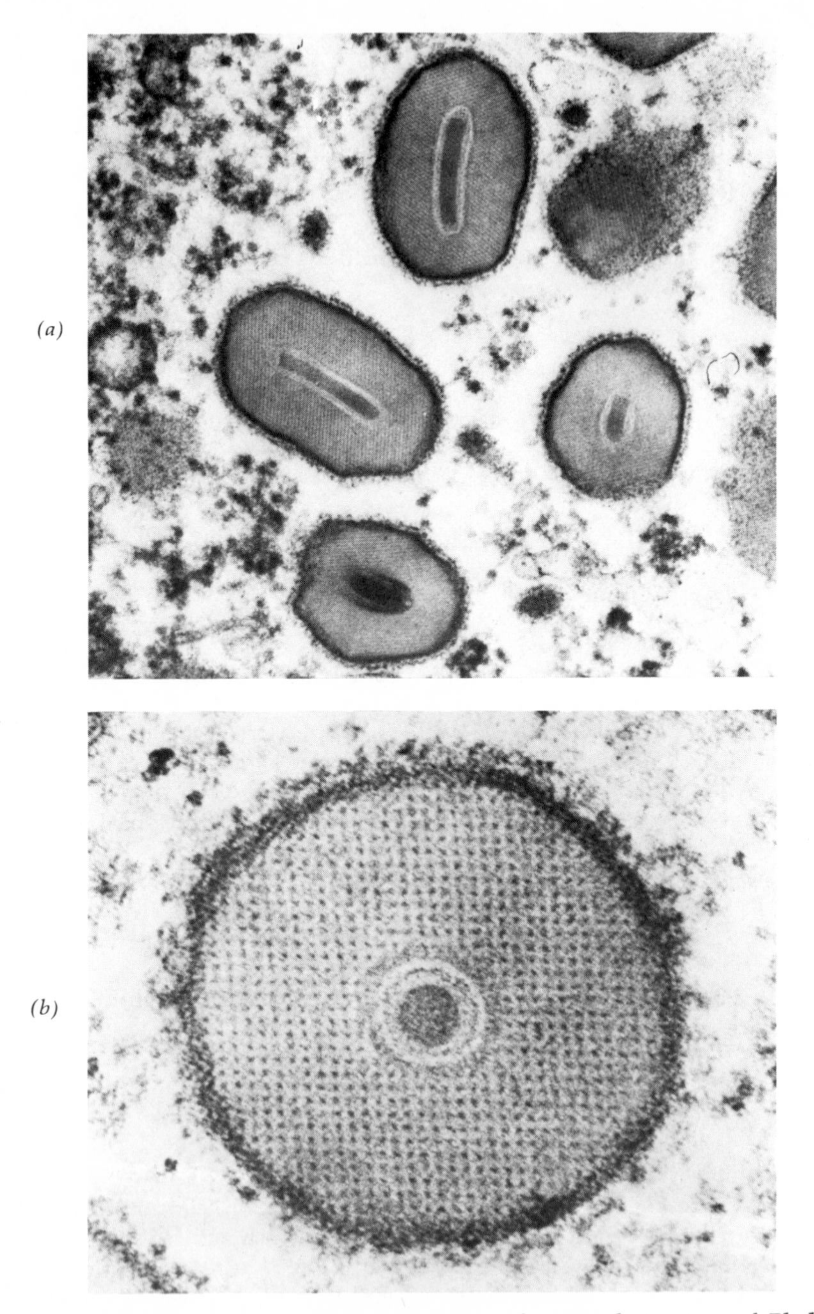

FIG. 3-18. **Sections of capsules containing the granulosis virus of Plodia interpunctella (a moth). Two virus rods are revealed in longitudinal section (a) and one in transverse section (b). Membranous structures appear to surround the centrally located virus rod and the crystalline lattice of the capsules is also apparent. (Courtesy of H. J. Arnott and K. M. Smith.)**

Melnick, J. L.: Classification and Nomenclature of Animal Viruses, *Prog. Med. Virol.* **13**:462–484, 1971.

Schäfer, W.: Some Aspects of Animal Virus Multiplication, *Perspect. Virol.* **1**:20–39, 1959.

Functions of Viral Constituents

Avery, O. T., C. M. MacLeod, and M. McCarty: Studies on the Chemical Nature of the Substance Inducing Transformation of Pneumococcal Types. Induction of Transformation by a Desoxyribonucleic Acid Fraction Isolated from Pneumococcus Type III, *J. Exp. Med.* **97**:137–157, 1944.

Fraenkel-Conrat, H.: The Role of the Nucleic Acid in the Reconstitution of Active Tobacco Mosaic Virus, *J. Am. Chem. Soc.* **78**:882, 1956.

Franklin, R. M.: The Significance of Lipids in Animal Viruses. An Essay on Virus Multiplication. *Prog. Med. Virol.* **4**:1–53, 1956.

Gierer, A., and G. Schramm: Infectivity of Ribonucleic Acid from Tobacco Mosaic Virus, *Nature* (London) **177**:702–703, 1956.

Hershey, A. D., and M. Chase: Independent Functions of Viral Protein and Nucleic Acid in Growth of Bacteriophage, *J. Gen. Physiol.* **36**:39–56, 1952.

Kates, M., A. C. Allison, D. A. J. Tyrrell, and A. T. James: Lipids of Influenza Virus and Their Relationship to Those of the Host Cell, *Biochim. Biophys. Acta* **52**:455–466, 1961.

Knight, C. A.: The Nature of Some of the Chemical Differences among Strains of Tobacco Mosaic Virus, *J. Biol. Chem.* **171**:297–308, 1947.

———: Nucleoproteins and Virus Activity, *Cold Spring Harbor Symp. Biol.* **12**:115–120, 1947.

Markham, R., and K. M. Smith: Studies on the Virus of Turnip Yellow Mosaic, *Parasitology* **39**:330–342, 1949.

Schäfer, W.: Some Aspects of Animal Virus Multiplication, in *Perspectives in Virology*, M. Pollard (ed.), pp. 20–39, Wiley, New York, 1959.

Steitz, J. A.: Identification of the A Protein as a Structural Component of Bacteriophage R17, *J. Mol. Biol.* **33**:923–936, 1968*a*.

———: Isolation of the A Protein from Bacteriophage R17, *J. Mol. Biol.* **33**:937–945, 1968*b*.

Gene Location

Davis, R. W., M. Simon, and N. Davidson: Electron Microscope Heteroduplex Methods for Mapping Regions of Base Sequence Homology in Nucleic Acids, *Methods Enzymol.* **21**:part D, 413–428, 1971.

Jeppesen, P. G. N., J. A. Steitz, R. F. Gesteland, and P. F. Spahr: Gene Order in the Bacteriophage R17 RNA: 5′-A Protein–Coat Protein–Synthetase-3′, *Nature* (London) **226**:230–237, 1970.

Kado, C. I., and C. A. Knight: Location of a Local Lesion Gene in Tobacco Mosaic Virus RNA, *Proc. Natl. Acad. Sci. U.S.A.* **55**:1276–1283, 1966.

Kado, C. I., and C. A. Knight: The Coat Protein Gene of Tobacco Mosaic Virus. I. Location of the Gene by Mixed Infection, *J. Mol. Biol.* **36**:15–23, 1968.

Morrow, J. F., and P. Berg: Cleavage of Simian Virus 40 DNA at a Unique Site by a Bacterial Restriction Enzyme, *Proc. Natl. Acad. Sci., U.S.A.* **69**:3365–3369, 1972.

Westmoreland, B. C., W. Szybalski, and H. Ris: Mapping of Deletions and Substitutions in Heteroduplex DNA Molecules of Bacteriophage Lambda by Electron Microscopy, *Science* **163**:1343–1348, 1969.

Morphology

Bradley, D. E., and D. Kay: The Fine Structure of Bacteriophages. *J. Gen. Microbiol.* **23**:553–563, 1960.

Brandes, J., and R. Bercks: Gross Morphology and Serology as a Basis for Classification of Elongated Plant Viruses, *Adv. Virus Res.* **11**:1–24, 1965.

Caspar, D. L., R. Dulbecco, A. Klug, A. Lwoff, M. P. G. Stoker, P. Tournier, and P. Wildy: Proposals, *Cold Spring Harbor Symp. Quant. Biol.* **27**:49, 1962.

——— and A. Klug: Physical Principles in the Construction of Regular Viruses, *Cold Spring Harbor Symp. Quant. Biol.* **27**:1–24, 1962.

Crick, F. H. C., and J. D. Watson: Virus Structure: General Principles, in *The Nature of Viruses*, G. E. W. Wolstenholme and E. C. P. Millar (eds.), pp. 5–13, J. & R. Churchill Ltd., London, 1957.

Finch, J. T., and K. C. Holmes: Structural Studies of Viruses, in *Methods in Virology*, K. Maramorosch and H. Koprowski (eds.), vol. 3, pp. 351–474, Academic, New York, 1967.

Horne, R. W.: The Structure of Viruses, *Sci. Am.* **208**:48–56, 1963.

Horne, R. W., and P. Wildy: Symmetry in Virus Architecture, *Virology* **15**:348–373, 1961.

Klug, A., and D. L. D. Caspar: The Structure of Small Viruses, *Adv. Virus Res.* **7**:225–325, 1960.

Wildy, P., and D. H. Watson: Electron Microscopic Studies on the Architecture of Animal Viruses, *Cold Spring Harbor Symp. Quant. Biol.* **27**:25–47, 1962.

Nucleic Acid Structure

Brownlee, G. G., and F. Sanger: Chromatography of ^{32}P-labeled Oligonucleotides on Thin Layers of DEAE-cellulose, *Eur. J. Biochem.* **11**:395–399, 1969.

———, ———, and B. G. Barrell: The Sequence of 5S Ribosomal Ribonucleic Acid, *J. Mol. Biol.* **34**:379–412, 1968.

Chargaff, E.: Structure and Function of Nucleic Acids as Cell Constituents, *Fed. Proc.* **10**:654–659, 1951.

Franklin, R. E., and R. Gosling: Molecular Configuration in Sodium Thymonucleate, *Nature* (London) **171**:740–741, 1953.

Kleinschmidt, A. K.: Monolayer Techniques in Electron Microscopy of Nucleic Acid Molecules, in *Methods in Enzymology* 12B, L. Grossman and K. Moldave (eds.), pp. 361–377, Academic, New York, 1968.

Robertson, H. D., and P. G. N. Jeppesen: Extent of Variation in Three Related Bacteriophage RNA Molecules, *J. Mol. Biol.* **68**:417–428, 1972.

Sanger, F., G. G. Brownlee, and B. G. Barrell: A Two-dimensional Fractionation Procedure for Radioactive Nucleotides, *J. Mol. Biol.* **13**:373–398, 1965.

Steitz, J. A.: Polypeptide Chain Initiation: Nucleotide Sequences of the Three Ribosomal Binding Sites in Bacteriophage R17 RNA, *Nature* (London) **224**:957–964, 1969.

Thomas, C. A., Jr., Ritchie, D. A., and L. A. MacHattie: The Natural History of Viruses as Suggested by the Structure of Their DNA Molecules, in *The Molecular Biology of Viruses*, J. S. Colter and W. Paranchych (eds.), pp. 9–30, Academic, New York, 1967.

Watson, J. D., and F. H. C. Crick: The Structure of DNA, *Cold Spring Harbor Symp. Quant. Biol.* 18, 123–131; *Nature* (London) **171**:737–738, 1953.

Wilkins, M. H. F., A. R. Stokes, and H. R. Wilson: Molecular Structure of Desoxypentose Nucleic Acids, *Nature* (London) **171**:738–740, 1953.

Williams, R. C.: Electron Microscopy of Sodium Desoxyribonucleate by Use of a New Freeze-drying Method, *Biochim. Biophys. Acta* **9**:237–239, 1952.

Protein Structure

Harris, I. J., and C. A. Knight: Studies on the Action of Carboxypeptidase on Tobacco Mosaic Virus, *J. Biol. Chem.* **214**:215–230, 1955.

Hennig, B., and H. G. Wittmann: Tobacco Mosaic Virus: Mutants and Strains, in *Principles and Techniques in Plant Virology*, C. I. Kado and H. O. Agrawal (eds.), pp. 546–594, Van Nostrand Reinhold, New York, 1972.

Schramm, G.: Über die Spaltung des Tabakmosaikvirus und die Wiedervereinigung der Spaltstücke zu höhemolekularen Proteinen, *Z. Naturforsch.* **2b**:112–121, 1947.

Sreenivasaya, M., and N. W. Pirie: The Disintegration of Tobacco Mosaic Virus Preparations with Sodium Dodecyl Sulphate, *Biochem. J.* **32**:1707–1710, 1938.

Stanley, W. M., and M. A. Lauffer: Disintegration of Tobacco Mosaic Virus in Urea Solutions, *Science* **89**:345–347, 1939.

CHAPTER 4
PROCESS OF INFECTION

Viruses are obligate parasites, which means that they must get into cells in order to multiply. In this they differ from pathogenic bacteria and fungi, which often multiply on the surfaces of tissues and in extracellular fluids as well as in cells. The problem, then, is how viruses of the diverse forms described in Chap. 3 get into appropriate cells and, once in, how they function.

The barrier to a virus, or to any foreign object, depends on the nature of the cell involved as well as on the size and character of the invading particle. Cells vary as much in size and form as virus particles; therefore, it is not possible to illustrate accurately all animal cells with a single example. This also holds for bacterial and plant cells. Nevertheless, some characteristic features of different types of cells are illustrated by the diagrams of Fig. 4-1.

First of all, it can be noted that all cells share some basic common features. The cellular contents in each case are contained within a sheath, the cell membrane, sometimes referred to as the *cytoplasmic* or *plasma membrane*, or *plasmalemma*. This membrane is composed mainly of lipids and polysaccharides combined with a variety of proteins, providing a constitution which apparently lends both strength and resilience to the structure. When ruptured, the membrane of a healthy cell is readily repaired by one or more of the enzymes present. The membrane shows a selective permeability for small molecules and ions and a capacity to engulf droplets and particles (*pinocytosis* and *phagocytosis*, respectively; *viropexis* when the particle engulfed is a virus). Conversely, the cell membrane may form filamentous protrusions, microvilli, the numbers of which are augmented by injury and certain chemicals, such as vitamin A. Both invaginations and protrusions are dynamic structures which can appear and disappear. The cell membrane is of paramount importance in the social behavior of

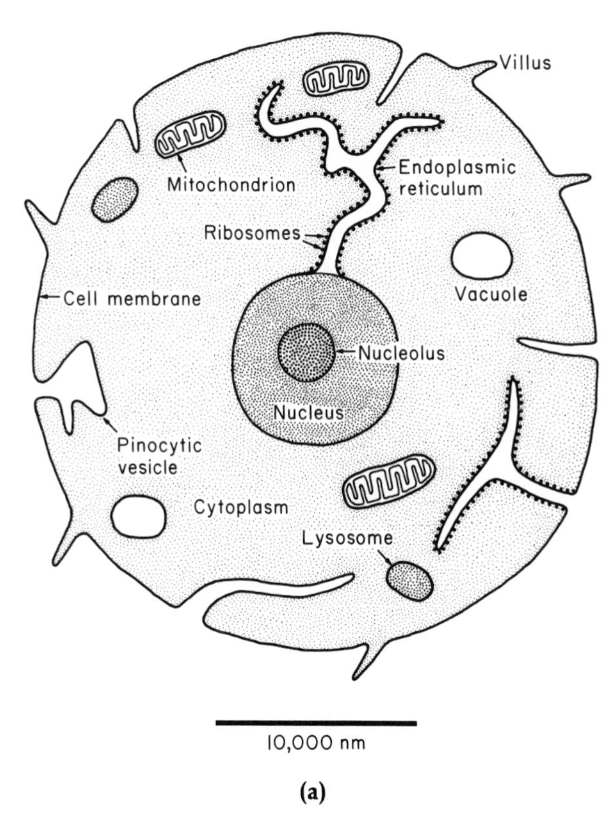

FIG. 4-1. **Schematic diagrams showing some features of the major types of cells infected by viruses: (a) animal cell; (b) bacterial cell; (c) plant cell.**

FIG. 4-1. **(continued on following page)**

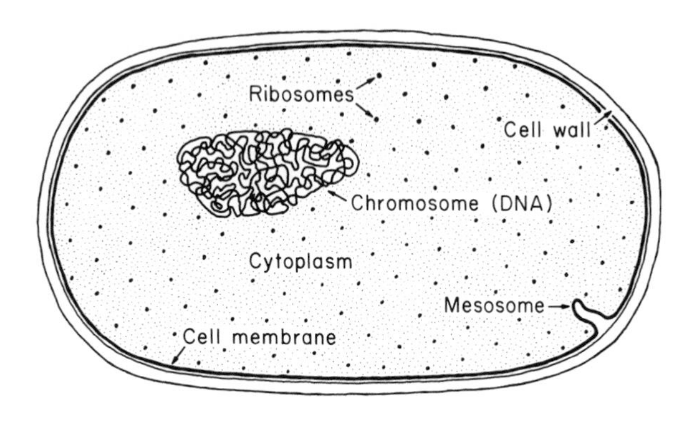
Ribosomes
Cell wall
Chromosome (DNA)
Cytoplasm
Mesosome
Cell membrane
1000 nm

(b)

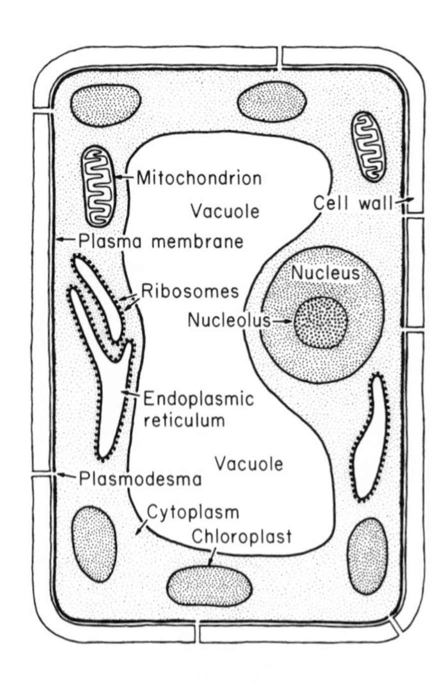
Mitochondrion
Vacuole
Cell wall
Plasma membrane
Nucleus
Ribosomes
Nucleolus
Endoplasmic reticulum
Vacuole
Plasmodesma
Cytoplasm
Chloroplast
10,000 nm

(c)

cells and in their immunologic reactions. Furthermore, in animal and plant cells, infoldings of the membrane are thought to constitute the convoluted network called the *endoplasmic reticulum* (in bacterial cells the comparable structures are called *mesosomes*) and perhaps the nuclear membrane as well. The vast surface areas and the chemical nature of membranes provide favorable sites for accumulating a variety of chemical substances, including enzymes and ribosomes, and consequently considerable metabolic activity occurs in such areas. Thus, the cell membrane is a very important common feature of bacterial, animal, and plant cells.

Another cardinal feature of the cells depicted in Fig. 4-1 is the presence of ribosomes. These organelles are composed of RNA and protein, and they constitute the machinery by means of which genetic messages are translated into gene products, i.e., into proteins.

The three types of cells considered here all have DNA as their primary hereditary material. They differ mainly in amount and in how it is packaged. The major DNA in nondividing animal and plant cells is found in a definite structure called the *nucleus,* whereas the DNA of bacterial cells is present as a mass in the cytoplasm, unbounded by a membranous structure. DNA is also present in such organelles as mitochondria, which are found in both animal and plant cells and also in chloroplasts of plant cells.

Turning to differences among cells, size is a major distinction when animal and plant cells are compared to bacterial cells. The former have of the order of a thousand times the mass of bacteria and a similar preponderance of DNA.

Another notable distinction is that bacterial and plant cells have cell walls in addition to cell membranes, whereas animal cells do not (Fig. 4-1). Such cell walls tend to be impervious to macromolecules, including virus particles, although it will be noted in the diagram of the plant cell that there are pores in the cell wall. These pores are called *plasmodesmata* when they communicate between adjacent cells, and *ectodesmata* when they communicate to the exterior of cells. Plant cells, it may be noted, often have rather large vacuoles compared with other types of cells.

The process of infection on the molecular level consists of the interaction of cells of the sort just sketched with viruses of diverse structure, as described in Chap. 3. The whole process can be briefly generalized for all viruses as follows. Virus attaches to a susceptible cell at more or less specific sites (a single infectious particle is sufficient to initiate infection). Either whole virus or viral nucleic acid penetrates to the interior of the cell. Multiplication of the virus takes place in either the cytoplasm or the nucleus, or in both. If whole virus has penetrated, multiplication is preceded by disruption of the virus particle in order to release the nucleic acid. The genetic information of the viral nucleic acid becomes functional through the same type of mechanism usually operative in cells, namely, viral DNA is transcribed into messenger RNA (mRNA), which is then translated into proteins. If the viral nucleic acid is RNA rather than DNA, it can act as mRNA directly, or in some cases (e.g., influenza and vesicular stomatitis viruses and viruses containing double-stranded RNA) the viral RNA is transcribed to provide mRNA. In either case, the proteins resulting from translation of mRNA are mainly enzymes needed in the synthesis of viral constituents and viral structural proteins. The host cell contributes heavily to the synthesis of a virus. Low molecular weight cellular building blocks, such as amino acids and nucleotides, are built into viral proteins and nucleic acid, respectively. The cell's ribosomes are employed in viral protein synthesis, and various enzymes and transfer RNAs (tRNAs) are contributed, although some new enzymes and tRNAs may be synthesized as specified by the virus. Viral macromolecules (mainly proteins and nucleic acid) accumulate in pools until a critical concentration is attained, after which assembly of progeny virus particles begins. However, with some RNA-containing viruses, it is possible that assembly is the terminal event of a transcription-translation-assembly sequence so that no pool of nucleic acid is actually ever present. Assembly can be partly or wholly a spontaneous process. Finally, mature virus particles are released from the infected cell ready to begin a new cycle of infection. Following a virus through its life cycle, the process of infection can be viewed as passing through the stages of attachment, penetration, multiplication, assembly, and release. Some distinctive features of each of these stages of infection will be considered next.

4-1 *Attachment*

THERE IS MUCH SPECIFICITY IN THIS STEP FOR BACTERIAL AND ANIMAL VIRUSES BUT LITTLE FOR PLANT VIRUSES

In systems in which bacterial and animal viruses are in a liquid medium containing susceptible cells, the virus particles are brought near to or in contact with the cells by diffusion and Brownian motion. Attachment (sometimes also called *adsorption*) leading to infection occurs upon collision of virus particle with cell surface if there is a structural and electrostatic complementarity between cell surface and virus particle. In short, there must be a fairly strong and specific affinity between viral attachment sites and cellular receptor sites. So far as is known, all the viral attachment sites as well as the cellular receptor sites are constituted at least in part of protein. The proteins may be conjugated, i.e., they may be glycoproteins or lipoproteins, especially in the case of bacterial cell walls or animal cell membranes. Proteins, including those of virus coats and of cell surfaces, have characteristic conformations based on their amino acid compositions. With these distinctive shapes go characteristic charge distributions related mainly to the free amino and carboxyl groups.

Emphasis is placed on electrostatic forces in the initial step of the infectious process because it has been observed that attachment occurs in most systems only at pH values where amino and carboxyl groups are largely ionized (pH 5 to 10), and selective destruction of amino or carboxyl groups on either viral or cell surfaces prevents attachment. Since attachment involves an interaction between charged particles, it would also be expected that this process would show sensitivity to the salt composition of the medium, and this is observed. Both cell surfaces and virus particles tend to have net negative charges over a wide pH range, and thus cells and viruses repel one another unless the ionic environment is modulated by the presence of salt ions. For example, several phages show very little attachment to bacterial cells if the salt concentration of the medium is 10^{-4} M or less, and the attachment of fowl plague virus or poliovirus to cells can be linearly correlated within certain limits with the logarithm of cation concentration.

Structural specificity in the attachment interaction between viruses and cells has been demonstrated in a variety of ways. Tailed phages of the T series attach to *Escherichia coli* bacterial cells by means of tail fibers. Mutations which affect the structure of these tail fibers can prevent such phages from infecting. On the other hand, the bacterial receptor structures are so selective for the phages which they will attach that it is possible to type various strains of bacteria on this basis. Another example is the sharp specificity that poliovirus shows for primate cells, which is apparently based on the affinity between viral attachment sites and receptor substances on the host cell membrane. However, this restrictive specificity disappears if viral RNA is the infecting species; infectivity then depends only on the capacity of the viral RNA to penetrate the cell membrane. Consequently, not only monkey and human cells are infected by poliovirus

RNA, but also cells of such nonprimate species as mouse, rabbit, guinea pig, chicken, and hamster. It should be noted that infection of nonprimate cells by poliovirus RNA does not result in spread of virus to uninfected cells, because the progeny virus of the initial infection have the usual protein coat, which does not permit attachment to nonprimate cell membranes. Another excellent example of the relationship between the virus coat protein and cell receptors in the process of infection is the demonstration that poliovirus RNA enclosed in the coat protein of Coxsackie virus infects mouse cells that are completely insusceptible to poliovirus RNA wrapped in its own coat protein (mouse cells are susceptible to infection by Coxsackie virus).

In contrast to the marked specificities of attachment leading to infection with animal and bacterial viruses, there seems to be no such parallel phenomenon with plant viruses. If the RNA of a strain of tobacco mosaic virus that is unable to infect tomato plants is used as inoculum rather than whole virus, no infection ensues. Moreover, coating this RNA (by reconstitution) with protein from a strain of TMV that does infect tomato plants fails to confer infectious capacity; conversely, coating the RNA of the strain which does infect tomato with the protein of the strain which does not fails to prevent the infection. Similar results have been observed with mixed reconstitution products of coat proteins and nucleic acids of the small spheroidal plant viruses, cowpea chlorotic mottle virus, brome mosaic virus, and broad-bean mottle virus. It seems that the specificity observed with plant viruses enters at some stage in the process of infection beyond the attachment and penetration steps.

4-2 *Penetration*

IN SOME CASES, VIRAL NUCLEIC ACID IS CARRIED INTO CELLS BY ENGULFMENT OF WHOLE VIRUS PARTICLES, WHEREAS IN OTHER CASES NUCLEIC ACID IS INJECTED INTO CELLS

The essence of infection by a virus is that the genetic activity of viral nucleic acid is superimposed upon or substituted for that of cellular nucleic acid. Therefore, in order to function it appears that the first step necessary is that a virus get its nucleic acid into the cell. This can be accomplished in various ways.

Many bacteriophages have evolved a mechanism which enables

them to inject their nucleic acid through the formidable barrier of the cell wall as well as through the contiguous cytoplasmic membrane.

Animal viruses seem able to penetrate readily those cells to which they attach; two modes of entry have been observed. Many animal viruses are engulfed as whole virus particles by a phagocytic kind of process sometimes called *viropexis*. The virus particles may be partly degraded in the penetration process either at the cell membrane, within phagocytic vesicles, or by means of lysosomes. The latter are organelles rich in enzymes which may merge with vesicles, by a process of membrane fusion. The other process, which has been observed especially with poliovirus (in addition to engulfment), involves passage of the virus through the cell membrane without benefit of a phagocytic mechanism. How this is accomplished is presently obscure, but the process is specific, since molecules considerably smaller than the virus are excluded.

Plant cells, as indicated in Fig. 4-1, have thick cell walls which bar passage of many materials. However, plant viruses succeed in penetrating here and there when applied mechanically to such organs as leaves or roots, because cell walls are leaky. Transient pores called *ectodesmata* protrude through the cell wall at intervals, and whole virus is apparently engulfed at these points. The engulfment of whole particles of tobacco mosaic virus has been clearly demonstrated with cultures of plant protoplasts. In addition, many plant viruses are inserted into cells fortuitously by insects in the course of feeding. Once inside a cell, a plant virus, or its nucleic acid, can pass to another cell via the plasmodesmata.

The three types of penetration just sketched are summarized in Fig. 4-2.

Thus by one means or another, viral nucleic acid gets into the interior of a cell. This is necessary for infection but not sufficient. For example, it is possible for whole virus to get into a cell but not to have its protein coat removed, and hence the viral nucleic acid cannot function. Or, if the system is one in which the viral nucleic acid is injected into the cell, the cell may contain especially active nucleases that destroy foreign (viral) nucleic acids before they can function significantly. If conditions permit survival of at least some viral nucleic acid, there still may be no virus multiplication for want of appropriate machinery or parts for fabricating new virus. Finally, as will be discussed in the next section, successful penetration of viral nucleic acid can lead to two different types of infection, one of which soon culminates in production of numerous new infectious viral particles while in another the production of viral progeny is delayed indefinitely.

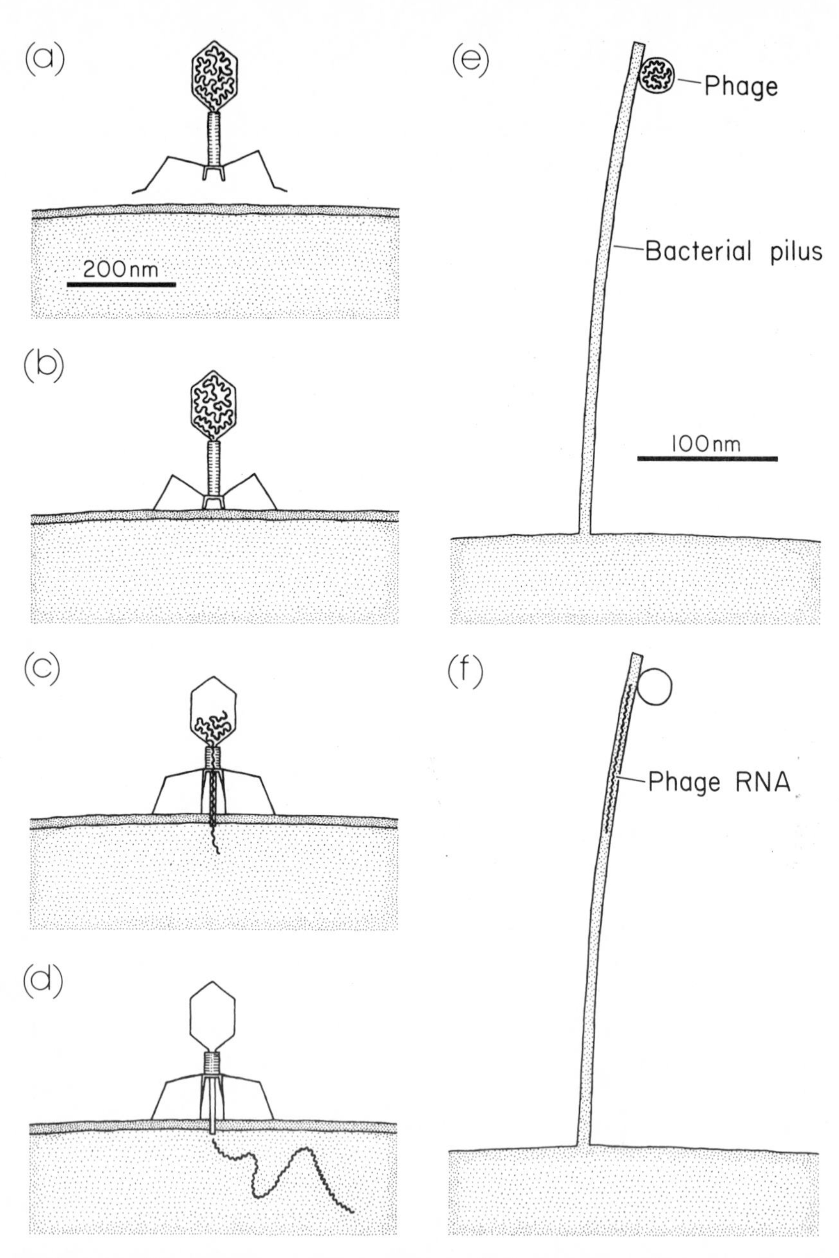

FIG. 4-2. **Diagrammatic representation of three types of penetration of cells by viruses. Type 1: (a) through (d) injection of DNA by a tailed bacteriophage; (e) and (f) injection of RNA into a bacterial pilus by a spheroidal bacteriophage. Type 2: (g) through (j) engulfment of a whole virus particle and its subsequent degradation in a cellular vesicle. Type 3: (k) and (l) direct penetration of a cell by a whole virus particle, mechanism obscure.**

FIG. 4-2. **(continued on following page)**

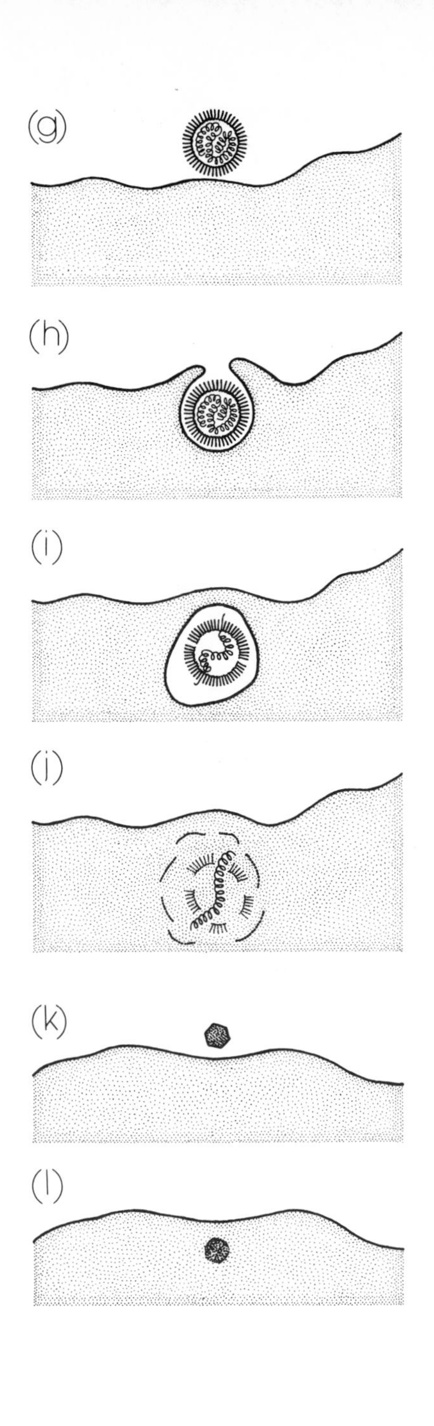

(g)
(h)
(i)
(j)
(k)
(l)

4-3 *Multiplication*

VIRAL NUCLEIC ACID AND PROTEIN ARE SYNTHESIZED MOSTLY BY MEANS OF HOST CELL MACHINERY FROM AVAILABLE CELL MATERIALS

One type of viral infection leads to production of hundreds or thousands of new viral particles per infected cell. The essence of this type of virus multiplication is twofold: replication of viral nucleic acid and production of a coat or coats to enshroud the nucleic acid.

Some preliminary arrangements are necessary before the synthetic apparatus of the cell is impressed into the manufacture of virus. Part of this adjustment involves changes to the virus, such as removal of protein coat, and part is concerned with synthesis of new enzymes or modification of old ones. In any case, shortly after attachment there is a period of time (which varies for different viruses or even for different virus-cell combinations) when assays which distinguish between whole virus and infectious nucleic acid show that there is no increase (often there is a decrease) in the kind of infectivity characteristic of whole viral particles. This is called the *latent period.* The very low level of infectivity demonstrable during the latent period is largely if not entirely attributable to a small portion of the inoculum which is not actively participating in the infectious process but just "sticking around." The particles of virus actively engaged in the process of infection are degraded ("eclipsed") during the latent period in order to release viral nucleic acid.

Similarities and differences between the three major types of viruses with respect to latent period are illustrated in Fig. 4-3. As the various curves indicate, there is no increase in measurable virus during the latent period, but the latter is followed by a rise period in which virus concentration increases markedly. Finally a plateau is reached, and the average number of infectious units per cell at this time is called the *burst size* for phages or just "the yield" for animal and plant viruses. It should be noted that the time scale for replication of phages is in minutes, whereas that of animal and plant viruses is in hours.

Synthesis of viral parts may take place in the cell nucleus, in the cytoplasm, or partly in both. In each case, the cell provides energy, virus precursor molecules, some of the enzymes which catalyze synthesis of viral parts, and other appropriate machinery. Cellular machinery for such syntheses seems to be especially active at membrane surfaces in either infected or uninfected cells. For example, the ribo-

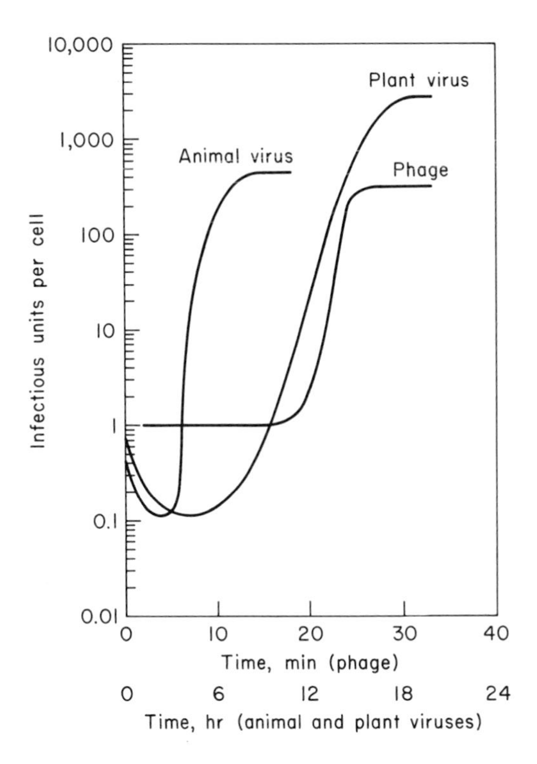

FIG. 4-3. **Curves illustrating the level of infectious virus at various times during one replicative cycle of infection. The curves, though not representing specific viruses, are typical of those observed for animal, plant, and bacterial viruses, respectively, in cell cultures.**

somes which constitute the structures essential to protein synthesis are often seen concentrated on the folds of membrane called *endoplasmic reticulum* (Fig. 4-1).

Viral nucleic acids are synthesized from constituent nucleotides (via nucleoside triphosphates), often employing replicase enzymes coded for by the nucleic acid of the infecting virus particle. The precise mechanism of synthesis of viral nucleic acid depends somewhat on the system and type of nucleic acid. Some modes of replication for which there is considerable experimental evidence are sketched in Fig. 4-4. Replication of double-stranded circular nucleic acid is not illustrated in the figure, but this presumably occurs by the semiconservative mechanism indicated in Fig. 4-4, an essential first step in the process being an enzymatic cut of at least one of the strands of the duplex. Another instance of the potential complexity of nucleic acid replication is that in some cases a bit of RNA appears to couple to one end of DNA where it acts as a primer for the replicase enzyme in starting the synthesis of more DNA. It should be noted that an important common feature of all modes of replication of nucleic acid is hydrogen bonding between complementary base pairs. Also it must be recalled that one or several enzymes are required in the process, such as a nuclease to open a strand in circular template DNA, polymerase(s) to hook nucleotides together in the growing polynucleotide chain, and sometimes a ligase to hook segments of polynucleotide together. There are also control mechanisms, at present ill-defined, which enable the preferential production of copies of one strand of nucleic acid rather than its complement. Thus by template mechanisms similar to those sketched in Fig. 4-4 and with the help of appropriate enzymes, one of the major functions of viral nucleic acid, namely, self-replication, is achieved. The resulting new strands of viral nucleic acid accumulate in the cell until their concentration and that of other viral components favor packaging the nucleic acid into a new virus particle. With such viruses as T2 and T4 bacteriophages, the pool size might reach something like 50 phage equivalents (a phage equivalent is the amount needed for one phage particle) of nucleic acid and structural proteins by the time the first infectious particles can be detected.

The second major function of viral genomes is coding for necessary proteins, either enzymes needed for the synthesis of virus constituents or viral structural proteins themselves. If the viral nucleic acid is DNA, the process of protein synthesis follows the sequence which has come to be known as the "central dogma" of molecular biology:

$$\text{DNA} \xrightarrow[\text{(RNA polymerase(s))}]{\text{"transcription"}} \text{Messenger-RNAs} \xrightarrow[\text{(ribosomes, tRNAs, etc.)}]{\text{"translation"}} \text{Proteins}$$

If the viral nucleic acid is double-stranded RNA, RNA substitutes for DNA in the central-dogma scheme. Single-stranded viral RNA often acts directly as mRNA, although such viral RNAs as those of influenza and vesicular stomatitis viruses have the wrong "polarity" (if the polarity of mRNA is designated plus, these viral RNAs have the complementary, or minus, polarity) and apparently must be transcribed to provide functioning mRNA. Another exception is seen with RNA tumor viruses; it has been proposed that in this case the RNA is first transcribed into DNA and then the usual sequence starting from DNA is followed.

All viral nucleic acids are polygenic, i.e., they code for several proteins. The simplest situation would presumably be a nucleic acid coding for only two proteins, a polymerase for replication of the nucleic acid and a coat protein. Most virus nucleic acids contain more messages than this, the number of proteins coded for varying roughly

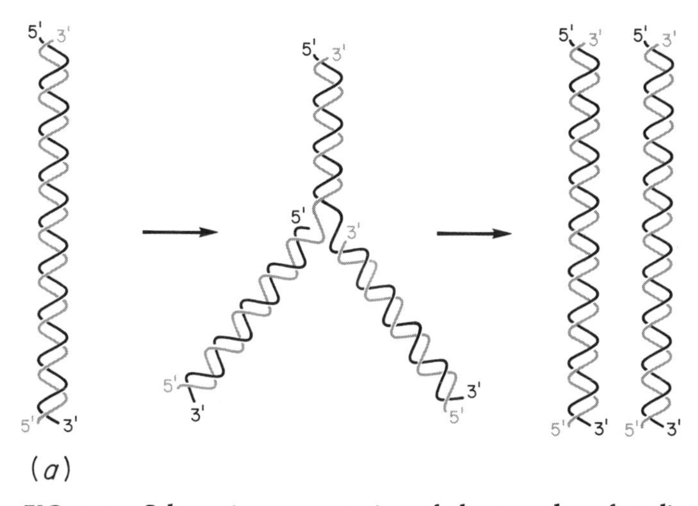

FIG. 4-4. **Schematic representation of three modes of replication of nucleic acids: (a) semiconservative replication of DNA. Note that each strand of the double helix serves as a template for replication of a new strand. Thus the resulting progeny duplexes contain one old strand and one new. (b) Conservative replication of double-stranded RNA. The RNA duplex opens up enough to permit the minus strand to serve as a template for synthesis of a new plus strand, thus forming a new duplex which can repeat the cycle. (c) Replication of single-stranded RNA. Viral plus strand serves as a template on which several minus strands can be made, and these in turn serve as templates for synthesis of more plus strands. Occasionally a double helical structure forms.**

FIG. 4-4. **(continued on following pages)**

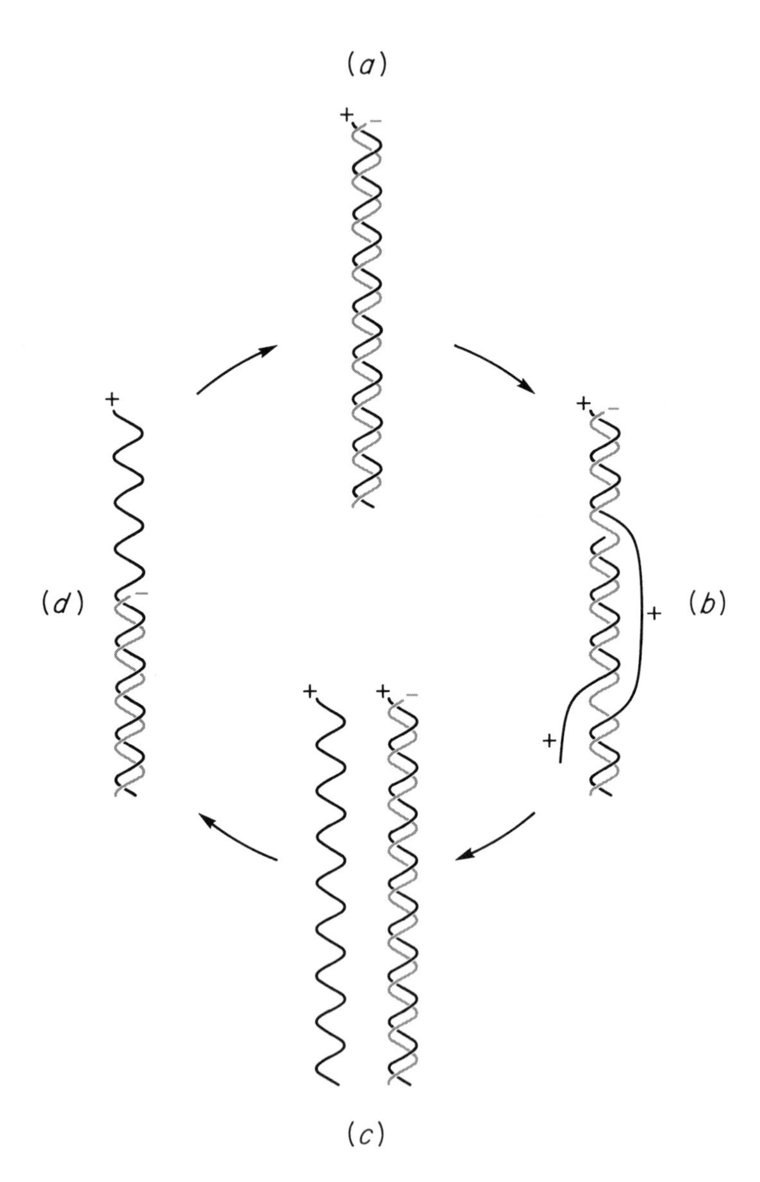

FIG. 4-4(b).

with the size of the nucleic acid. The simplest RNA-containing viruses, such as R17, f2, and Qβ phages and brome mosaic virus, may have only three or four genes; tobacco mosaic virus, poliovirus, and small DNA phages such as ϕX174 may have half a dozen or so; while

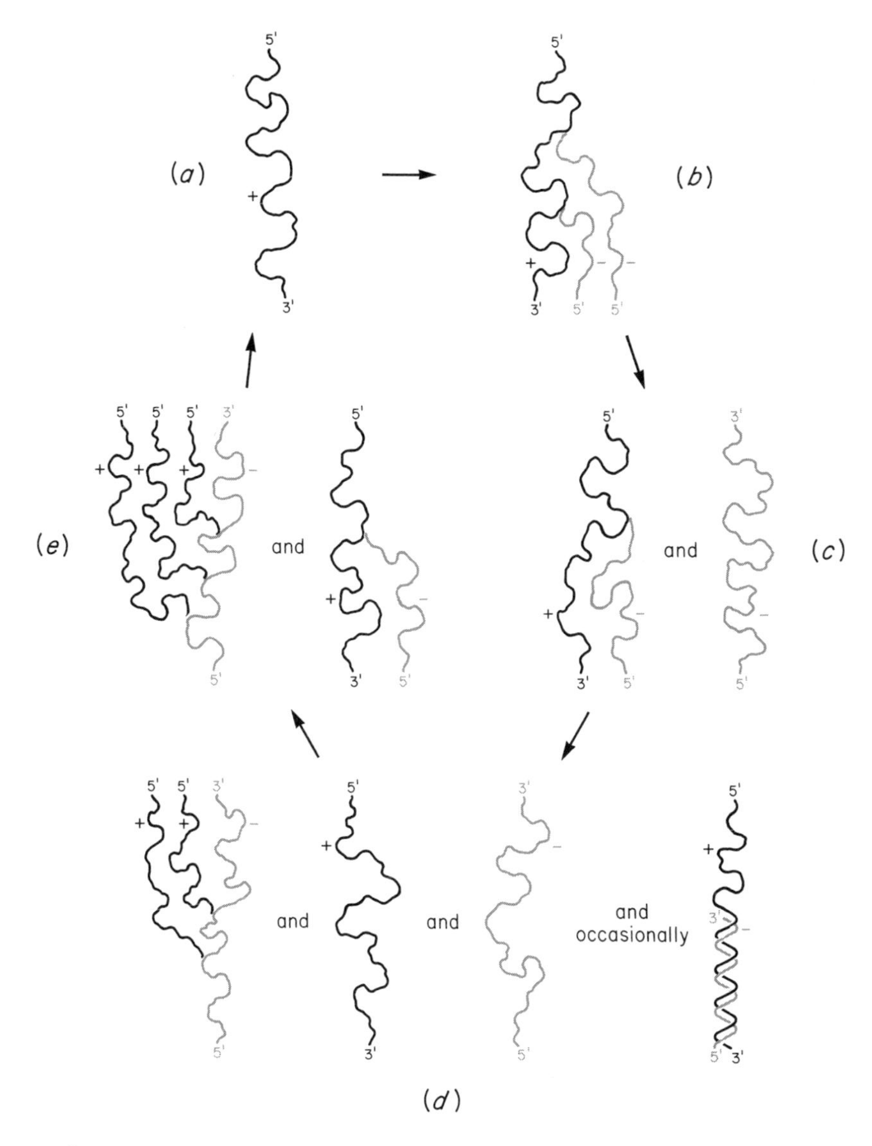

FIG. 4-4(c).

large phages such as T2 and T4 and the poxviruses have enough nucleic acid to code for 200 to 300 proteins and hence could have a similar number of genes. There are hints that not all the viral nucleic acid codes for proteins; some portion of it, for example, may serve as an attachment ("recognition") site for the replicase enzyme which catalyzes synthesis of new viral nucleic acid off the template of the old. Furthermore, in the case of double-stranded nucleic acids, it appears that usually only one of the two strands is transcribed into mRNAs and hence into proteins. T4 and lambda phages are exceptions, one strand of the DNA being transcribed for certain genes and the other strand for the remaining ones. Similarly, some genes are transcribed from one strand and some from the other of the *E. coli* bacterial chromosome.

In any case, enzymes which are needed for viral replication but which are unavailable in the host, and viral structural proteins are coded for by the viral nucleic acid. In addition, it seems that in the more complex viruses there may also be proteins that inhibit combination of host cell mRNAs with ribosomes; proteins which serve as initiation factors in the replication, transcription, and translation of viral nucleic acid; and proteins concerned with the morphogenesis and maturation of viral particles. Viral enzymes and proteins are synthesized on ribosomes by the common cellular mechanism sketched in bare detail in Fig. 4-5. As mentioned previously, functioning ribosomes are commonly found attached to cell membranes. Furthermore, one consequence of infection seen with some viruses such as poliovirus is that a viral protein renders cellular mRNAs incapable of attaching new ribosomes, thus accumulating these for viral protein synthesis.

Two distinct modes of viral protein production have been observed. A common one leads to production of individual species of viral proteins in a temporal sequence. In the other instance, exemplified by poliovirus and Coxsackie virus, the entire nucleic acid of the virus (these viruses contain a single strand of RNA) is translated to produce a single polypeptide chain whose molecular weight is over 200,000. This large polypeptide chain is then apparently cleaved by proteolytic enzymes at specific points to yield enzymes and structural proteins needed for building new virus particles. This is "processing."

Various control mechanisms appear to operate in regulating the production of viral proteins. In the case of some viruses, such as the T4 coliphage, control appears to be exercised at the transcription level. Immediately after infection, a few viral genes are transcribed by a host cell RNA polymerase whose activity is apparently circumscribed by its ability to initiate transcription at a limited number of sites on the phage genome. A short time later this polymerase is chemically altered, perhaps by replacement of the host initiation factor with a virus-specified one which permits it to transcribe a

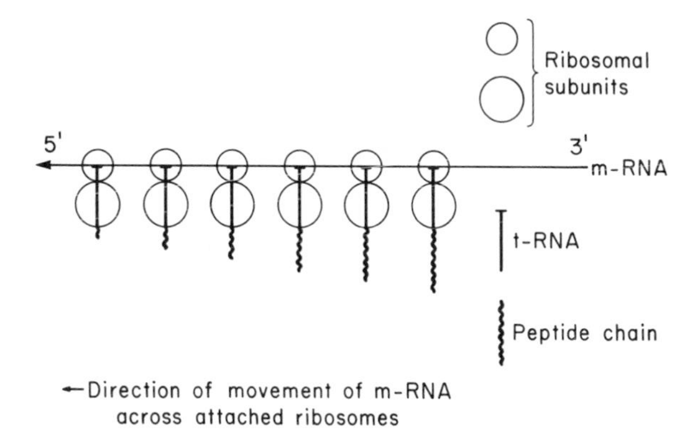

FIG. 4-5. **A polyribosome functioning in protein synthesis. The preface to the sketch shown here includes the following: Messenger RNA (mRNA) couples at a particular point to a small ribosomal subunit, to which a larger subunit then joins. A transfer RNA (tRNA) molecule, to which a specific amino acid is attached, hydrogen bonds through its anticodon triplet of nucleotides to the complementary mRNA triplet (codon) near the 5′ terminus of the RNA strand. Similar attachments of tRNAs with their amino acids to adjacent codons of the mRNA occur, and an enzyme (peptidyl transferase) located on the larger ribosomal subunit joins amino acids by forming peptide bonds between them. The mRNA moves across the ribosome, and more ribosomal subunits attach to form a polyribosome, and the process of peptide chain initiation is repeated successively on the new ribosomes. Chain initiation is followed by chain growth through the linking of more amino acids in the same manner. In the sketch above, this process has progressed to the point where a completed peptide chain has been released from the distal (3′) end of the mRNA, together with tRNA and the now-separated ribosomal subunits. Nascent peptide chains in the process of elongating are shown on the various ribosomes. For simplification, the enzymatic steps involved in coupling amino acids to tRNA and the signals required for initiation and termination of the peptide chain have been omitted. For details of these phenomena see references at the end of the chapter.**

second group of mRNAs. The mRNAs produced by the first two rounds of transcription are translated to produce "early" proteins, many of which appear to be enzymes concerned with synthesis of new viral DNA. Upon synthesis of progeny viral DNA, transcription of the genes coding for early proteins is largely "switched off," while genes for "late" proteins (largely viral structural proteins) are turned on. This is associated with a further modification of the RNA polymerase and with the presence of single-stranded regions in nascent viral DNA. The latter may serve as new initiation signals.

A variation of this type of control is seen with coliphage T7, whose genes concerned with early functions are transcribed by a host cell polymerase, whereas the late genes are transcribed by a phage-coded polymerase.

By contrast a significant level of control of synthesis of vaccinia viral proteins appears to occur at the translational level. As with other viruses, the proteins synthesized early in the replicative cycle of vaccinia virus are mainly enzymes needed in the replication of the virus, especially of the viral DNA. There is a dramatic switch-off of the synthesis of early proteins at a certain point in the viral replicative cycle, and this switch-off appears connected with viral DNA replication and protein synthesis. If these latter activities are inhibited by appropriate metabolic antagonists, switch-off does not occur on schedule. Therefore, it is concluded that one of the first late proteins to be synthesized selectively prevents the translation of the mRNAs for early enzymes.

The small, spheroidal, RNA-containing phages of the f2 class exhibit another example of control of protein synthesis at the translational level. The genomes of these phages contain only three genes, which, starting at the 5′ terminus of the RNA, occur in the order A protein, coat protein, and RNA polymerase (see Chap. 3). The viral RNA has plus polarity and hence acts directly as mRNA. If each gene were translated in sequence, equal amounts of A protein, coat protein, and RNA polymerase would be produced. This is not the case. Large numbers of coat-protein molecules are synthesized but only few A-protein and polymerase molecules. Concomitant with this disproportionate synthesis of viral proteins, the viral RNA must serve as a template for its own replication, by the mechanism illustrated in Fig. 4-4*c*. Thus transcription and unequal numbers of translational events involve the same RNA molecules, and the RNA must be attached to ribosomes for translation but must be free of ribosomes for transcription. All this must be neatly regulated to produce phage efficiently, and these phages do multiply very well. The programming of multiplication appears to proceed as follows. There are three sites, rather than one, at which the phage RNA can be attached to ribosomes for translation, one at each of the three genes. However, the phage RNA, although mainly single-stranded, has some regions where base complementarity permits formation of double-stranded loops. A consequence of this structure is that not all ribosome attachment sites are equally accessible, the coat-protein site being most favorable. Hence much more coat protein than A protein or polymerase is made. Furthermore, coat protein appears to bind to both the A-protein and polymerase genes in a way that prevents attachment to ribosomes at those sites. Hence as coat protein accumulates, translation of A-protein and polymerase genes is progressively more inhibited. The interference between transcription and translation is thought to be surmounted by the ability of polymerase (which works from 3′ to 5′ on the RNA, whereas translation by ribosomes goes from 5′ to 3′) to attach to the ribosome-binding site of the coat-protein gene as well as to the 3′ terminus of the viral RNA. When incumbent ribosomes have left the region between these points, the RNA polymerase proceeds to transcribe the RNA (assuming that the A-protein gene is only seldom occupied by ribosomes).

These examples suggest that a variety of controls can function in the regulation of viral protein synthesis, and the plant virus system has not yet been explored in this regard. Elucidation of these phenomena seems likely to have considerable significance, not only for virology but also in understanding gene expression in general.

The productive type of virus multiplication just described occurs sooner or later with virtually every virus. However, there are important instances when the production of virus components is put off indefinitely. This arrangement of deferred multiplication is called *lysogeny* in the phage-bacterial cell relationship and *virogeny* for viruses in general. Virogeny is a well-established phenomenon with bacterial viruses; it is less well established with animal viruses and has not yet been reported for plant viruses.

A cardinal feature of virogeny is that the nucleic acid of the invading virus becomes integrated into host cell nucleic acid, where, instead of multiplying extensively and at its own pace as it does in productive virus multiplication, it is duplicated only when host cell nucleic acid is duplicated prior to cell division. In the integrated state, bacterial viral nucleic acid is referred to as *prophage;* in the case of animal viruses and viruses in general the term used is *provirus.* The productive and lysogenic cycles of replication of such a virus as a coliphage lambda are schematically illustrated in Fig. 4-6.

Some major principles of the deferred virus production characteristic of the lysogenic cycle in bacteria, and which probably apply more or less to virogeny in general, may be summarized as follows. After the viral nucleic acid is released in a cell, circumstances must favor its integration into host DNA; otherwise the viral nucleic acid becomes operative, as discussed previously, in production of progeny virus. Some factors which determine whether integration of viral nucleic acid will occur are the genetic constitutions of the virus and of the host cell, multiplicity of infection, nutritional state of the host cell, and temperature. There are some viruses which seem hereditarily incapable of virogeny and likewise some cells which seldom if ever allow integration of viral nucleic acid. With phages, at least, a high multiplicity of infection (e.g., 10 infectious particles per cell), a low temperature (20°C rather than 37°C), and a starved nutritional state of the cell favor lysogeny. It should be noted also that an essential condition for virogeny is that a virus contain double-stranded DNA or the capacity to form it within a cell.

In virogeny, it is clear that there must be some means of integrating viral nucleic acid into host nucleic acid and of maintaining it there. Successful integration of a nucleic acid such as that of phage lambda appears to employ a genetic recombination mechanism which

FIG. 4-6. **Lytic and lysogenic cycles of coliphage lambda schematically illustrated. The drawing is not to scale, the phage particles and parts being shown many times larger than they should be in proportion to the size of the bacterial cell. At the top of the sketch for the lytic cycle a bacterial cell is outlined within which the bacterial chromosome (DNA) is represented in a circular form. Proceeding clockwise, a phage particle is shown attached tail-first to a bacterial cell, and next the phage DNA has been injected into the cell, where it is represented by a wiggly line. Multiplication of phage DNA (wiggly lines) and phage proteins (irregular dots) are shown in the next cell, followed by assembly of some of the phage parts to produce incomplete and complete phage particles. Finally, a burst cell is shown from which phage parts and complete phage particles are being released, the latter to initiate a new cycle of infection. It will be noted that the bacterial chromosome is shown intact up to the lysis stage; the cycle for virulent phages such as T2 and T4 would resemble this except that the bacterial chromosome is degraded shortly after injection of the phage DNA.**

In the lysogenic cycle, the phage DNA (wiggly line) is shown integrated into the bacterial chromosome and duplicating with it to produce one copy in each daughter cell. However, upon induction, as shown in the sketch, phage parts are produced in abundance, and the following events are exactly those of the lytic cycle.

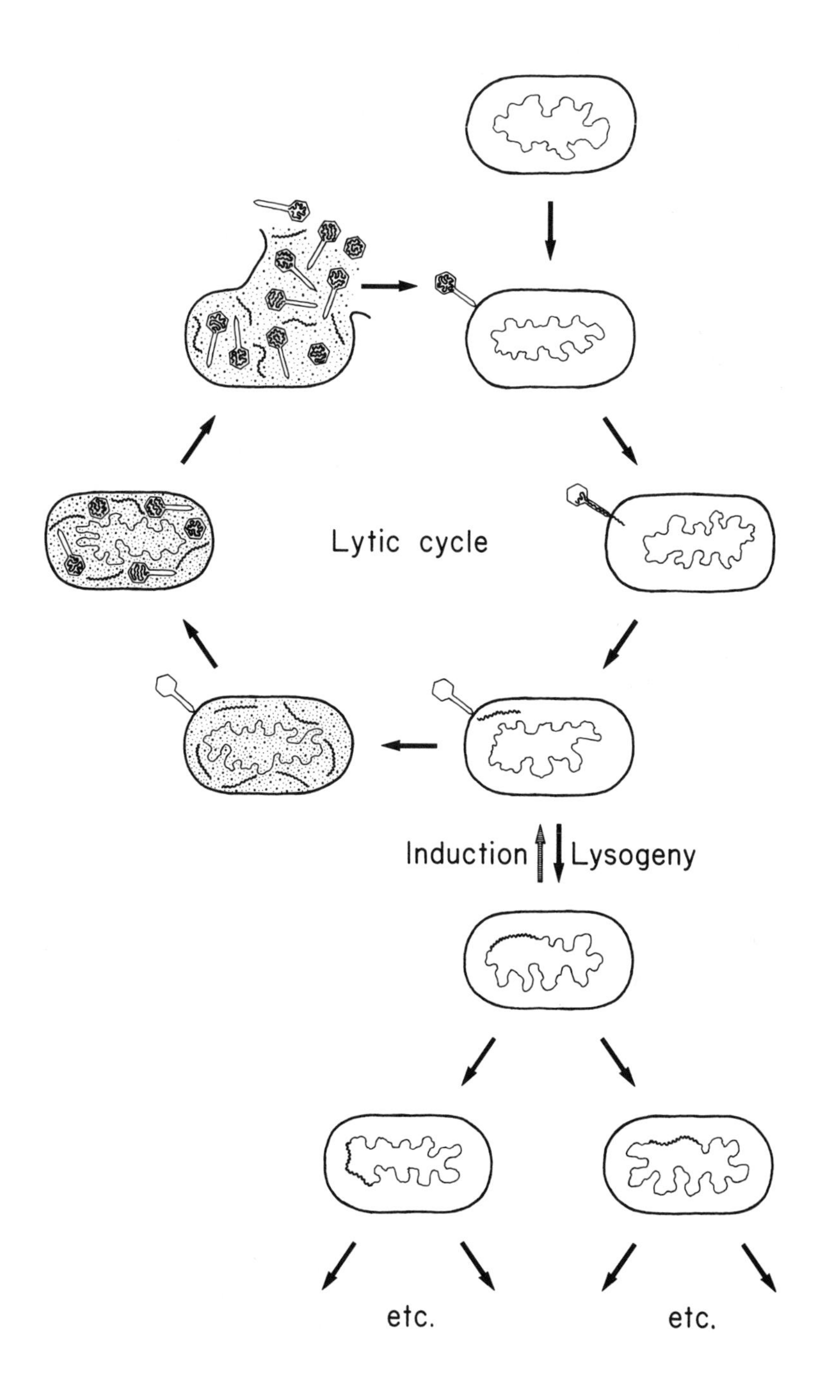
Lytic cycle
Induction
Lysogeny
etc.
etc.

requires attachment of viral nucleic acid to some specific site on the host nucleic acid, followed by opening of circles of nucleic acid and subsequent formation of covalent bonds (integrative recombination) between open ends of the nucleic acid chains. The reversal of integration, resulting in release of viral nucleic acid from host nucleic acid (called *induction*), requires virtually the same steps as integration. Both make use of specific enzymes, at least some of which are coded for by viral nucleic acid. Possible steps in integration and induction are shown diagrammatically in Fig. 4-7.

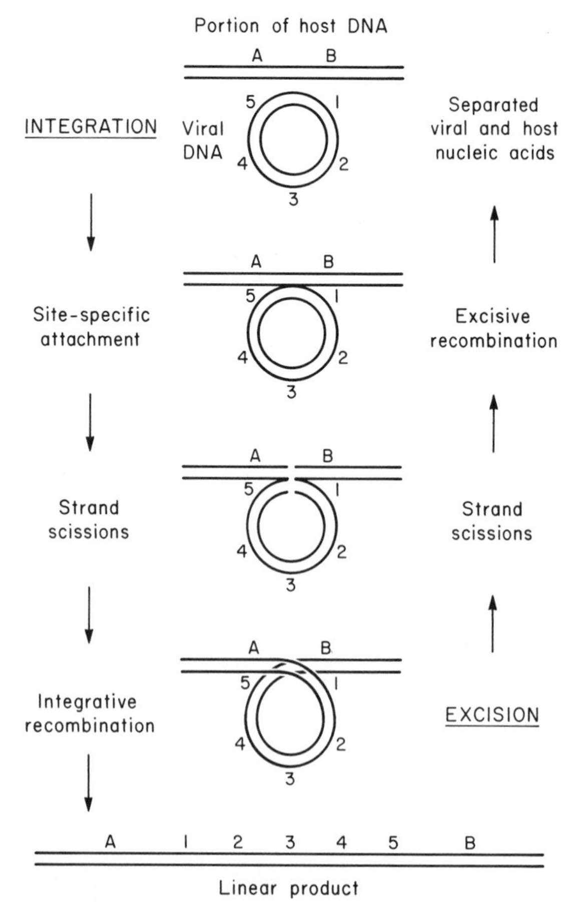

FIG. 4-7. **Possible mechanisms for the integration of viral DNA into and excision from host DNA in virogeny.**

In the virogenic state, most of the genes of the provirus are inactive. This inactivity is maintained in bacterial lysogeny by the functioning of one or a few viral genes, especially a gene which codes for a repressor protein. The repressor protein attaches to a key point on the viral DNA (operator region) and thus prevents the transcription of a series of genes needed for production of phage. Occasionally (about once in 10^{-2} to 10^{-5} cell generations) spontaneous induction occurs and the cell goes from the lysogenic into the lytic cycle, as indicated in Fig. 4-6. Induction seems to be influenced by some of the same factors as integration. The genetic constitution of the phage is crucial, whereas that of the host cell is relatively unimportant. This is illustrated by the observation that in the same strain of *E. coli* bacteria, phage lambda is readily inducible, while phage P2 is non-inducible. Richly nourished cells induce readily; starved cells do not. Experimentally, a wide variety of inducing agents has been found. These include radiations such as ultraviolet light, organic peroxides, nitrogen mustard, ethylene imine, fluorouracil deoxyriboside (FUDR), and mitomycin C. It is not clear how these diverse agents cause induction. Some may act directly to inactivate the repressor protein or the gene producing it, but considerable evidence points to indirect inactivation of repressor as a secondary effect or to a more subtle mechanism relating to the general health and repairability of the host DNA into which the viral genome is integrated. For example, the lag period between inducing treatment and release of repression is many times the time required for direct inactivation of repressor protein, and several inducing agents are ones known to interfere with DNA synthesis and/or repair rather than to affect proteins or protein synthesis directly. Among the bizarre circumstances causing induction are those in which the DNA-containing integrated provirus is brought into a strange cell. This can happen during conjugation of bacterial cells when a lysogenized donor cell transfers a phage-containing chromosome into a nonlysogenic cell (zygotic induction). Similarly, mouse fibroblast cells virogenized with simian virus 40 (SV40) can be induced by fusing such cells with monkey kidney cells.

In any case, when induction occurs, viral propagation goes forward in the manner characteristic of the particular virus involved, except as noted next.

Induction of a prophage is accompanied by release of the viral DNA from host DNA, into which it has been integrated. Infectious phage results, which is usually just like the phage originally infecting the bacterial cell. If, however, the excision of phage DNA is imprecise, part of the phage DNA may be left behind and a corresponding piece of bacterial DNA removed. This combination DNA incorporated into a phage particle is defective with respect to ability to support subsequent replication of the phage, because phage genes were left behind. On the other hand, the phage in initiating a new infection carries some bacterial genes into another cell. This process, discovered and elucidated by Joshua Lederberg and his associates, is called *transduction*. Transduction means "a leading across," and in the instance just described some bacterial genes are led across from one cell to another by a virus.

There appear to be two distinctive types of transduction: restricted (or specialized) and general. Restricted transduction is exemplified by coliphage lambda in its lysogenic relationship with *E. coli* K12. Lambda DNA appears from genetic mapping experiments to be integrated into the K12 DNA between the genes involved in utilization of galactose (the *gal* genes) and those concerned with biotin synthesis (*bio* genes). If lambda phages are obtained by inducing a lysogenized strain of K12 bacteria which has the capacity to ferment galactose (gal^+) and then are used to infect bacteria which cannot ferment galactose (gal^-), about 10^{-6} of the gal^- bacteria appear to acquire the gal^+ character. Likewise the transduction of *bio* genes by lambda has been observed. Since lambda can transfer only the genes next to it, i.e., *gal* or *bio* genes, it is said to function in restricted transduction. Other phages which attach to and integrate at other positions on the bacterial chromosome can transduce other bacterial genes in the same restricted manner.

In general transduction, any bacterial gene can be transferred by a phage particle from one bacterial cell to another. This occurs at a low frequency, about 10^{-5}, and differs markedly from restricted transduction in that little or none of the phage DNA may be involved. The process thus appears to resemble phenotypic mixing (see Chap. 5), in which a piece of bacterial DNA is packaged in a phage head along with or instead of the phage DNA. Another major difference between this process and restricted transduction is that such pickups of bacterial genes can occur not only upon induction of a lysogenized bacterium but also in lytic infections.

Both types of transduction represent intriguing methods by means of which a virus can act as a vector in transferring nonviral genes from one cell to another.

Virogeny (or "temperate infection" as contrasted to lytic infection) is of special interest in animal virology for several reasons, among which is the possibility that transformation of cells to a tumorous state may be associated with the state of the viral nucleic acid in this relationship.

Four major groups of DNA-containing viruses and one group of large, enveloped, RNA-containing viruses can currently be linked to oncogenesis, i.e., tumor generation. For short, these are often called DNA tumor viruses and RNA tumor viruses.

The DNA tumor viruses include the papovaviruses (exemplified by rabbit papilloma virus, simian virus 40, and mouse polyoma virus), adenoviruses, herpesviruses, and poxviruses. The best-known RNA tumor viruses are those of chicken, mice, hamsters, and cats.

Not all members of these groups are tumorigenic, and even with the tumorigenic viruses this quality varies with the host involved. For example, none of the human adenoviruses is known to cause tumors in man. However, human adenoviruses 12, 18, and 31 are said to be highly oncogenic in newborn hamsters, whereas numerous other adenoviruses (distinguished by serologic reactions) are either weakly oncogenic or not oncogenic at all in this host. Similarly, SV40 causes a lytic infection of monkey cells but a temperate (potentially tumorous) infection of mouse cells, while polyoma virus causes a lytic infection of mouse cells but a temperate infection of rat or hamster cells.

In temperate infection with DNA tumor viruses little or no infectious virus is produced, nor can infectious DNA be extracted from the cells. Nevertheless, viral genome must be present, since viral mRNA and virus-specific antigens are produced; furthermore, cells temperately infected with SV40 can be induced, with a resultant yield of whole infectious virus. Similarly, certain cells suspected of harboring RNA tumor viruses although not displaying any, do yield, upon subjection to inducing treatments, particles closely resembling RNA tumor viruses. These and other observations suggest by analogy with lysogeny that tumor virus genomes may be integrated into host DNA (in RNA tumor viruses, the viral genome must first be transcribed into double-stranded DNA by a "reverse transcriptase," as discussed under Virogenes in Chap. 7). Evidence supporting integration, especially in the cases of SV40 and polyoma virus DNAs, was obtained by hybridizing cellular DNA with RNA complementary to viral DNA. The viral specific RNA was made in vitro by *E. coli* RNA polymerase using viral nucleic acid as template. Thus, while the case for integration of tumor virus genomes is not as strong as that for prophage, it seems a good working assumption that it exists.

If integration of tumor virus genomes does occur, it may be

argued that there should be similarities between temperate infections of bacteria and of animal cells. This seems to be the case.

In both instances the genetic constitutions of virus and host cell are involved, so that the virogenic relationship is permitted in some cases and not in others.

High multiplicity of infection seems to favor virogeny in both systems.

The DNA tumor viruses resemble temperate phages in that in temperate infection most viral genes are turned off. In lysogeny, a few phage genes function to produce repressor which prevents other genes from functioning; it is not clear yet whether or not repressor is produced by tumor viruses, but certain genes presumably do function to produce virus-specific antigens (e.g., the ones called *tumor* or *T antigen* and *transplantation antigen*).

In lysogeny, it appears that one copy of phage DNA is integrated per bacterial chromosome; in DNA tumor viruses the figure may be about the same or less (estimates from hybridization experiments range from 1 to 64 equivalents of SV40 DNA per cell).

Lysogenic bacteria, as previously noted, are spontaneously induced at a low rate to produce infectious virus, and this rate of induction can be substantially increased in some cases, but not in all, by radiations or by treatment with certain chemicals; the same is true for some animal tumor viruses, e.g., SV40-transformed cells produce traces of infectious virus (spontaneous induction?) and can be induced to produce much more virus by ultraviolet irradiation or by short treatment with substances that interfere with DNA replication. On the other hand, polyoma virus apparently cannot be induced, just as some temperate phages cannot be induced. A suggested mechanism for this failure of induction is that in both cases incomplete viral genome has been integrated.

Transfer by cell conjugation of a bacterial chromosome containing a phage genome from a lysogenized cell to a nonlysogenic cell results in production of infectious virus ("zygotic induction"); similarly, if a nonpermissive SV40-transformed cell (a nonpermissive cell generally permits temperate but not lytic infection) is fused with a permissive cell, infectious virus is produced by the fused cell (i.e., the "heterokaryon").

Changes in antigens of the bacterial cell wall have been observed as a consequence of lysogeny (this is an example of lysogenic or phage conversion). If the transplantation antigen characteristic of animal tumor viruses is virus-coded, as is generally assumed, then the change which this antigen causes in cell surface may be considered a comparable effect.

Clearly there are common features of virogeny in bacterial and animal cells, and the general phenomenon of deferred multiplication of viruses represents an intimate and intricate association between viral and host genomes.

4-4 *Assembly*

ASSEMBLY OF VIRUS PARTICLES FROM COMPONENT PARTS MAY BE WHOLLY OR PARTLY A SPONTANEOUS PROCESS

Evidence accumulated over the years indicates that the major constituents of viruses, such as the protein subunits and nucleic acid, are not linked by covalent bonds. This suggests that the assembly of virus particles might be a spontaneous process. This assumption gained support when Fraenkel-Conrat and Williams showed that tobacco mosaic virus could be reconstituted from its protein and nucleic acid constituents just by mixing them in a test tube in dilute salt at about pH 7 (see Reconstitution, Chap. 7). The low molecular weight components of the virus were found to combine in a matter of minutes to form characteristic high molecular weight, rod-shaped particles possessing high infectivity. Such reconstituted virus appears virtually indistinguishable from native virus when tested for nuclease resistance and examined for structural details by means of electron microscopy and x-ray diffraction. Subsequently, it was found that such spheroidal plant viruses as cowpea chlorotic mottle virus, broad-bean mottle virus, and brome mosaic virus could be reconstituted readily in the laboratory from component parts. As might be imagined, in vitro reconstitution of the more complicated multiprotein viruses is less easily accomplished. However, partial success has also been obtained with poliovirus and with several phages, lending support to the notion that self-assembly may occur even with the most complex virus particles. On the other hand, the work of Wood, Edgar, and associates linking the assembly of the multipart T4 phage with a series of viral genes suggests that occasionally a step in the morphogenesis of a complex virus particle may require a special function. For example, an enzyme specified by T4 gene 63 seems to be involved in the attachment of tail fibers to the phage particle.

4-5 Release

VIRUSES LEAVE CELLS BY TRAVERSING SPECIAL CHANNELS, BY EXTRUSION OR BUDDING THROUGH MEMBRANES AND CELL WALLS, OR BY EXPULSION THROUGH GAPS IN RUPTURED CELLS

There seem to be practical limits to the amount of virus that will accumulate in an infected cell. In the extreme case, cells burst, releasing virus particles and other cell contents into the surrounding area. This is the characteristic finale of the lytic type of infection of bacteria by virulent phages. Such irreversible rupture appears to be aided by two phage-coded enzymes, a lipase which attacks the cell membrane and a lysozyme that hydrolyzes cell wall. Once these barriers are breached, the cell bursts, a victim of osmotic forces. The autolysis ("cytocidal effect") observed in the infection of cells by some animal viruses such as poliovirus, mengovirus, and Newcastle disease virus resembles the lytic effect observed with some phages. However, this kind of release of animal viruses may represent a rather unspecific event occurring late in the cycle of infection and reflecting the dying state of the cell, a condition which allows many constituents, including viruses and viral parts, to be released to the exterior. Cellular autolytic enzymes doubtless destroy cell membrane in moribund cells, although the possibility that a viral enzyme might play a role cannot be excluded in all cases.

Earlier in the infectious cycle, virus particles may accumulate in vesicles or cisternae, some of which are connected by tubules to the exterior of the cell. That virus is released through such tubules is strongly suggested by coupling the electron microscopic observations of virus in cisternae and in tubules with the demonstration rather early in the cycle of infection of infectious virus in the medium surrounding the infected cells. Thus the two mechanisms by means of which a virus such as poliovirus is released from an infected cell include the just-postulated tubular passage over an extended period of time and the final burst type of release from a dying cell. The movement of plant viruses from infected cells probably resembles the tubular type of exit, the channels in this case occurring in the plasmodesmatal pores (Fig. 4-1) which communicate between cells.

Extrusion or budding is another release mechanism exhibited by viruses. This has been observed with the fibrous DNA-containing phages of *E. coli* such as f1, fd, and M13 (there are also mutant strains of these phages which are released by lysis) and also with many animal viruses. The release of influenza and of Rous sarcoma viruses by budding differs significantly from the phage extrusion in that the animal viruses in budding out seem to acquire a portion of host cell membrane in the process, whereas the phages do not. Moreover, the budding of enveloped animal viruses represents a stage in the assembly and maturation of these viruses. Though the nucleoprotein portion of enveloped viruses is made in the interior of the cell, the several components of the envelope appear to be concentrated at the cell membrane, where they are fabricated into the virus particles just as they bud. Presumably the same kind of mechanism operates in the cases of enveloped plant viruses such as potato yellow dwarf and tomato spotted wilt viruses. Just as the engulfment of virus particles does little permanent damage to cell membranes, so does the reverse, the budding of viruses. Apparently the cell membrane is readily repaired in a viable cell and can withstand the exit of hundreds of virus particles.

The yield of virus particles released per cell varies, as might be expected, with the virus, the type of cell, and growth conditions. The average yield of virus particles per cell for bacterial viruses (often called the *burst size*) can range from 10 or 20 to 1,000. Often the yield is a few hundred. The yield of plant and animal viruses may be one to several orders of magnitude larger than that of bacterial viruses, ranging from a few thousand to a million particles per cell.

4-6 *Viral Pathogenesis*

BOTH ACTIVE AND INACTIVE VIRUS CAN AFFECT CELL PHYSIOLOGY IN CHARACTERISTIC WAYS. SOME OF THE CONSEQUENCES INCLUDE ABERRATIONS OF NUCLEI AND THEIR CONTENTS, CROWDING DERANGEMENTS, ALTERATIONS IN STRUCTURE AND FUNCTION OF CELL MEMBRANES, AND VARIOUS TOXIC EFFECTS

It has been apparent for many years that viruses can cause pathologic effects on their hosts, although the range of disease symptoms observed varies from nothing to massive destruction of infected cells. Among the most conspicuous cytopathic effects is death of cells, or cytocidal effect. This terminal, destructive consequence of virus infection, which is central in the concern of mankind about virus diseases, has been invaluable in the quantitative assays of viruses of many kinds. Groups of killed cells (called *plaques* in bacterial and animal cells and *local lesions* on plant leaves) can be regularly obtained experimentally (Fig. 4-8), and since their numbers are proportional to the amount of infectious virus present, constitute a precise measure of infectivity. Despite their great usefulness, such lesions are gross effects of viral infection. The application of microscopy, including electron microscopy, and of biochemical techniques has revealed finer details of viral pathogenesis, although much remains to be explained.

Tissue cultures of cells respond to infection with many animal viruses in a characteristic manner. The cells, which may initially be somewhat elongated, round up and look denser, i.e., show increased refractility. Microscopic examination often reveals nuclear condensation (pycnosis), perhaps accompanied by deformation of the nucleus, e.g., from a spheroidal to a kidney shape. DNA-containing material, which stains characteristically and hence is called *chromatin,* tends to be pushed to the periphery of the nucleus (*margination*). Chromosome damage has been observed with some slowly replicating viruses such as measles and mumps. In the course of lytic infection of *E. coli* bacteria by some virulent T phages, as followed by electron microscopy, centrally distributed *E. coli* DNA moves early in the infection toward the cell walls and shortly disappears (destroyed by viral-coded deoxyribonuclease).

A few enveloped animal viruses such as Sendai virus or SV5 (either active virus or ultraviolet-inactivated) cause cell membranes to fuse, enclosing masses of cytoplasm and a few to hundreds of

FIG. 4-8. **Illustration of plaques and spots obtained with some viruses. Although hundreds of cells constitute the area of one plaque (or necrotic spot), each originated from a single infected cell. (a) Plaques caused by phage P22 on a lawn of** *Salmonella typhimurium* **bacteria. The culture is on nutrient agar in a petri dish. The plaques represent groups of lysed cells. Note a few small clear plaques indicative of plaque-type mutants. (Courtesy of J. R. Roth.) (b) Plaques of Newcastle disease virus formed on monolayers of chick fibroblast cells attached to one side of a prescription bottle. After culture under a layer of agar, the overlay is removed and the sheet of cells is fixed and stained with an alcoholic crystal violet solution, rinsed, and air-dried. The background of uninfected cells retains stain, whereas groups of infected cells, which are grossly degraded, appear clear. (Courtesy of F. L. Schaffer.) (c) Local lesions on a leaf of tobacco (***Nicotiana tabacum* **L. cv. Xanthi-nc), caused by infection with a strain of tobacco mosaic virus. Each necrotic spot represents a group of dead cells, and the virus is confined to such lesions.**

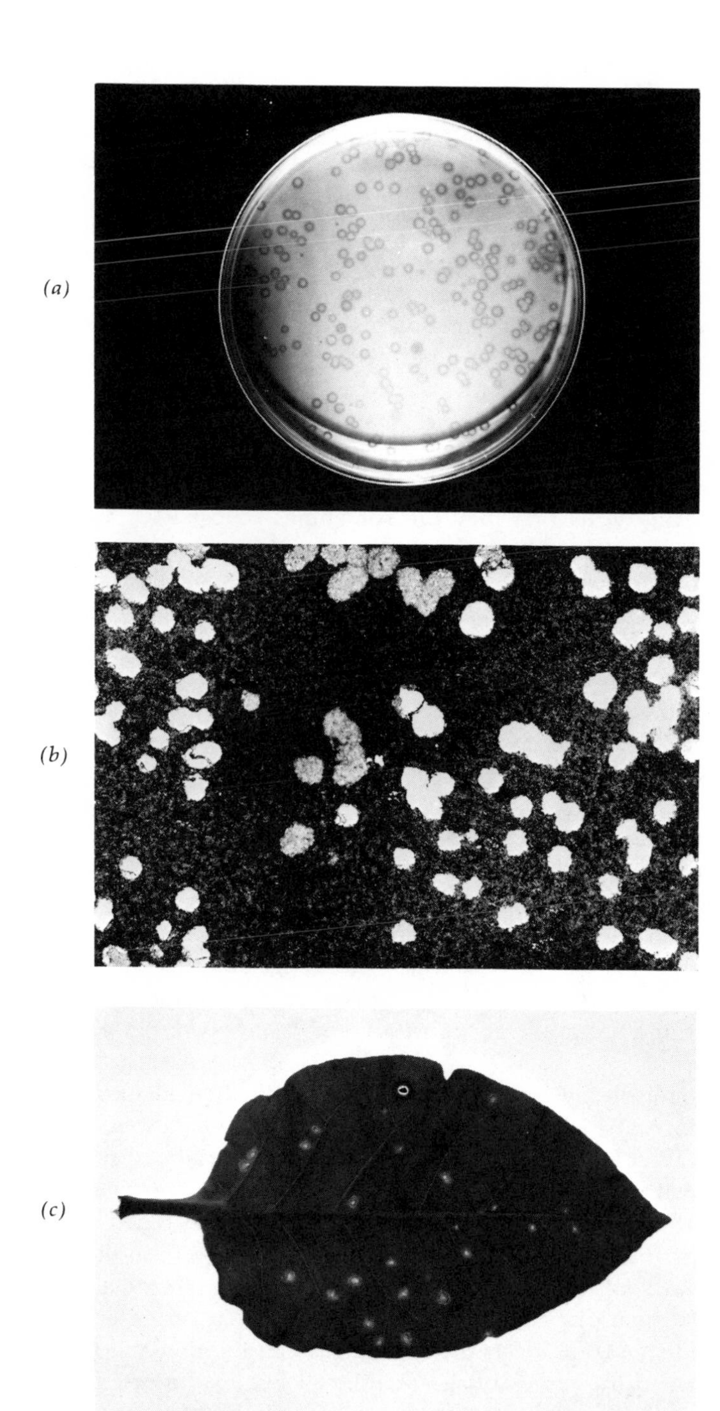

(a)
(b)
(c)

nuclei (such conglomerate structures are called *syncytia*). Syncytia formation is possibly mediated by fusion of viral envelopes with cell membranes. Aside from its interest as a dramatic example of viral pathologic effect, the phenomenon of cell fusion provides a remarkably useful method for mixing cytoplasmic and nuclear contents of a variety of cells. As mentioned elsewhere, viruses which do not induce cell fusion may significantly alter cell membranes by concentration of viral proteins and other constituents there. This conceivably affects transport of materials in and out of the cell.

Granular or fibrillar masses are often observed in the cytoplasm of virus-infected cells, whether they are bacterial, plant, or animal cells. These masses have been termed *inclusion bodies*. They occur in various sizes and shapes, which are so dependent on the nature of the infecting virus that they can sometimes be used for diagnosis of a disease. For example, Negri bodies constitute the most distinctive histopathologic feature of rabies. Some inclusion bodies are crystalline, and some are amorphous. Curiously, three-dimensional crystals of tobacco mosaic virus can be found in infected plant cells, whereas needle-like paracrystals (two-dimensional crystals) are more commonly obtained in the laboratory (Fig. 1-6*a*).(The needle-like crystals and other inclusions are also found in mosaic-diseased plants.) The cytoplasmic inclusions induced by infection with poxviruses often consist of a matrix containing viral DNA, viral protein, and some virus particles. Since at least some viral parts are made and assembled in such matrices, they have been referred to as "virus factories." Other viruses appear to cause similar inclusions either in the cytoplasm or in the nucleus, including putative virus factories observed in cells infected with TMV. It has been suggested that the sheer mass of inclusions in some cases may displace organelles and interfere with function and intracellular transport and hence contribute to pathogenesis.

Several of the consequences of viral infection mentioned above could contribute to the death of cells (necrosis) when that is the outcome of infection. In addition, there is the possibility of major interference with the synthesis of macromolecules and organelles needed, because these cellular constituents are normally degraded with time. Following the early lead of S. S. Cohen and colleagues in their studies of the biochemical consequences of infection by phages, it has been found with various animal systems that infection can interfere selectively with several aspects of the cellular transcription and translation processes, to the disadvantage of the cell but to the proliferation of the virus. Thus, cellular transcription may be inhibited by viral attack on host DNA (an extreme case is the drastic degradation of *E. coli* DNA by the nucleases of certain T phages) or the viral alteration of RNA polymerases. Some viruses, such as polio- and herpesviruses, interfere at the translational level by causing a disaggregation of polysomes engaged in host syntheses. The mechanisms of these inhibitions are generally unclear, but it appears that complete replication of virus, or even any replication of virus, may not be necessary. In fact, it has been known for years that T2 coliphage ghosts, i.e., phage particles devoid of DNA, prevent bacteria from multiplying and also cause their lysis. Similarly, just a few heat-inactivated vaccinia virus particles are enough to kill a cell, and vesicular stomatitis virus inactivated by ultraviolet light can prevent cell division. It appears possible that viral structural proteins might be responsible for some of these pathologic effects of viruses. The demonstrated cytotoxicity of one of the proteins of the adenovirus particle supports this idea. However, much more remains to be learned in this area.

4-7 *Control of Virus Infections*

VIRUS INFECTIONS GENERALLY RUN THEIR COURSES WITH VARIOUS DEGREES OF RESTRAINT; THE APPLICATION OF VACCINES AND ANTIVIRAL DRUGS ENHANCES THE RESTRAINT OF ANIMAL VIRUSES

It appears that all forms of life are so constituted as to resist destruction by viruses, although infected individuals may not thrive as well as uninfected ones. Even where virus infection may be highly destructive to a group, there are usually some individuals which are genetically more resistant than others and which selectively survive. Such natural resistance to viruses is deliberately fostered by man in some cases, e.g., by crossbreeding of susceptible plants with resistant ones. A difficulty is that viruses can mutate in a manner which may enable them to overcome host resistance.

In many animals two natural types of resistance to viruses exist which seem not to have counterparts in plants or bacteria. They are the production of antibodies and of a substance called *interferon*.

It has been known for many years in human medicine that recovery from such virus diseases as smallpox and measles left people immune to further attack for a few years or even for a lifetime. Such immunity seems to be correlated with the presence of humoral and tissue antibodies able to neutralize the virus. It will be noted that this natural immunity comes only after one has suffered through the disease. Such immunity is good, but obviously prophylaxis would be better. Consequently, the practice of vaccination of men and other animals, started by the English physician Edward Jenner in 1796, has gradually developed over the years. Vaccination consists of injection intradermally or intramuscularly of one or several doses of virus. Sometimes a mild ("attenuated" or "modified") strain of virus is employed in the vaccine. This mild strain of virus usually causes a localized infection which elicits the formation of antibodies directed against the virus in the vaccine and, most importantly, also against nonattenuated, virulent strains of the virus. Thus Jenner used cowpox virus to cause a mild, localized infection which resulted in the production of antibodies against cowpox but which also were effective against the related, virulent smallpox virus. Similarly, in more modern times mild strains of poliovirus are taken orally (Sabin vaccine) in order to initiate mild infections of the human intestine, which in turn cause production of antibodies against paralytic strains of poliovirus as well as against the mild vaccine strains. In other instances, as in the case of influenza, vaccination is done with inactivated viruses. Viruses can be rendered noninfectious but still antigenic (able to induce formation of antibodies) by judicious treatment with such agents as formaldehyde, ethylene oxide, β-propiolactone, or ultraviolet light.

The efficacy of a vaccine depends somewhat on the nature of the virus involved, its concentration and its antigenicity, and on the individual receiving the vaccine. There is a significant difference in the response of different individuals to the same amount of the same antigen.

In 1957 Isaacs and Lindenmann reported that infection with influenza virus caused cells to produce and release a soluble inhibitor which protected uninfected cells against infection. This inhibitor was named *interferon*. Now it seems that there is a series of interferons, depending on the type of cell in which they are produced. Human interferon is not the same as chicken interferon, which is not the same as horse interferon, etc. Furthermore, not only certain viruses but also many bacteria, bacterial endotoxins, acidic polysaccharides, and double-stranded RNA have all been found to induce cells to produce interferon. Double-stranded RNA, either natural or synthetic, is perhaps the most potent of these inducers. Inhibitors of protein synthesis tend to block the production of interferon with viral inducers but not with other inducers. This observation has led to the hypothesis that viruses cause synthesis of interferon, whereas other agents may just effect its release from a supply of interferon already present in cells. Interferon seems to be a protein, perhaps a glycoprotein, which is fairly heat resistant and stable over a wide pH range. Viral-induced interferons have a molecular weight of about 30,000.

Virtually all types of cells from all types of vertebrates, including tortoises and fish, have the capacity to produce interferons. Within a given type of cell, interferon shows no specificity. For example, interferon induced by a given strain of influenza virus in chick cells is effective against various types and strains of influenza virus but is also effective against unrelated viruses such as poxviruses and encephaloviruses. However, interferons do show a high degree of host cell specificity. Chick cell interferon is much more effective in chick cells than in calf cells, and horse interferon is almost useless in man.

The mechanism of action of interferon is not yet clear. It has no demonstrable effect on cell growth or metabolism, nor does it affect the penetration and uncoating of the virus. Considerable evidence indicates that interferon is not directly an inhibitor of virus replication but rather that it induces a second protein, coded for by the host genome, which does the inhibiting. In this scheme, interferon derepresses a gene in the host DNA from which mRNA for an inhibitory protein (also called *antiviral protein*) is transcribed. This antiviral protein subsequently inhibits translation of viral mRNA but not host cell mRNA. However, there is also evidence that antiviral protein acts earlier in the infectious cycle, namely, that by reaction with virus-specific RNA polymerase it prevents transcription of viral nucleic acid. Possibly both these actions may be attributable to the antiviral protein induced by interferon.

It is speculated that interferon serves to reduce the multiplication of animal viruses prior to the time that antibody can be developed. Judging from the severity of many virus diseases, this mechanism cannot be very efficient; and unhappily, mild strains of viruses seem to be better interferon inducers than severe strains. Moreover, as mentioned above, homologous interferon is required, so that it is impractical to produce large amounts of interferon in one host for use in another. At present a hopeful prospect in human treatment is that potent noninfectious inducers of interferon can be administered prophylactically so that production of interferon can be internally enhanced. Synthetic double-stranded ribopolynucleotides are good interferon inducers, but they also tend to be rather toxic in doses large enough to be effective inducers. New, potent inducers with only slightly toxic effects may ultimately be found, but it will be difficult to surpass natural double-stranded RNAs in these regards. The route of administration may be crucial in determining the practicality of any inducer. For example, in experiments with mice treated with the RNA preparation called statolon (obtained from *Penicillium stoloniferum*), it was found that intraperitoneal injection, although causing high levels of interferon in the blood, afforded little protection against infection with influenza virus. In contrast, statolon administered intranasally prior to infection provided protection lasting for days. Apparently a barrier prevents circulating interferon from reaching the surface cells of the respiratory tract where influenzal infection begins. The results of such experiments, combined with the knowledge that the intranasal route is a common portal of entry for animal viruses, suggest that the intranasal administration of interferon inducers may prove to be a convenient and efficient technique for protecting against respiratory viral infection.

The success of antibiotics against bacterial infections fostered the hope that similarly efficacious inhibitors of viral infections would be found. This hope has been only partly realized. Two factors work against chemotherapy of viruses: (1) though there are many chemicals which readily inactivate viruses or prevent their replication, they are so generally toxic to cells that the cure may be worse than the disease; (2) timing of administration is often unfavorable. For example, in many cases virus multiplication is almost over by the time characteristic symptoms appear and medication is begun; in cases where the virus multiplies more slowly diagnosis is often difficult. The infection may prove to be bacterial rather than viral. Nevertheless, some antiviral drugs show promise at least in limited applications, and the rationale for developing antiviral chemotherapy is clear, namely, to find substances which interfere specifically with viral protein or nucleic acid synthesis. Since it is already known that the replication of some viruses involves enzymes which do not exist in uninfected cells, one approach is to find specific inhibitors for these enzymes. Another is to enhance host cell functions (as with Isoprinosine) so that they suppress viral syntheses.

The multiplication of viruses which contain DNA is inhibited by halogenated pyrimidines such as 5-iodo-2′-deoxyuridine (IUDR) (Fig. 4-9*a*) or the analogous bromo- and fluoro- compounds. These substances are incorporated into newly synthesized DNA in place of thymidine where they may represent lethal mutations or cause faulty transcription. Thus the control of DNA viruses can be quite good but tends to be complicated by the general toxic effect of these halogenated pyrimidines. Excellent results have been obtained with IUDR in the control of viral eye infections of man, such as those caused

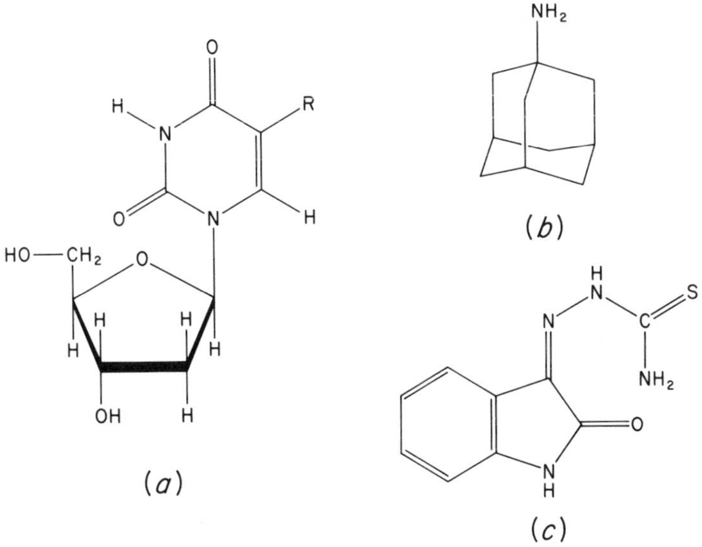

FIG. 4-9. **Structural formulas for some antiviral drugs. (a) Halogen-substituted deoxyuridine. R can be iodine, bromine, or fluorine. (b) Amantadine (*l*-adamantanamine). (c) Isatin-β-thiosemicarbazone (IBT).**

by herpes simplex virus. Some purine derivatives also show promise as antiviral agents. One of these is Isoprinosine, an alkyl amino alcohol complex of inosine. Its precise mode of action is not known, but, since it induces an increase in the rate of synthesis of mammalian mRNA in several tissues in vivo, it has been suggested that it enhances the capacity of host mRNAs to compete with viral mRNAs for host ribosomes on a mass action basis. There is also some evidence that the drug may alter polyribosome conformation and function in such a way as to induce the ribosomes to reject viral mRNA in favor of host messenger. In any case, Isoprinosine was found to exert antiviral effects against influenza virus and herpes-, polio-, and adenoviruses in tissue culture systems and to show therapeutic antiviral effects against influenza and herpes infections in newborn mice. Encouraging results have also been obtained in initial human tests.

Amantadine (*l*-adamantanamine hydrochloride) (Fig. 4-9*b*) and some of its derivatives appear to be effective against certain enveloped viruses such as influenza virus. These substances are better prophylactic than therapeutic agents, probably because their primary action is to interfere with penetration of the virus.

Thiosemicarbazones seem to interfere with the replication of poxviruses but not of most other viruses. One of these compounds, isatin-β-thiosemicarbazone (IBT) (Fig. 4-9*c*), has been found to block translation of the mRNA for late viral proteins, which include most of the coat proteins of the virus. Hence no progeny virus is formed. In human tests, IBT appears to be effective only if given in advance of infection and has the unfortunate side effect of causing severe vomiting.

In general, the results obtained with antiviral drugs are sufficiently positive to encourage the search for or synthesis of new ones. An ideal antiviral drug would of course interfere specifically with one of the steps of viral replication discussed earlier in this chapter without serious toxic reactions or side effects. Hopefully, it would also be effective against a wide variety of viruses.

The virus controls based on neutralizing antibodies and interferons are related almost exclusively to animal viruses, although Ross and others have evidence that plants may produce interfering substances as a consequence of infection with plant viruses (see Interference, Chap. 5). As in the case of animal interferons, the plant substances are of dubious significance in altering the course of natural infections.

Antiviral chemicals hold some promise for control of plant virus diseases as they do for other types of viral infections, but at present the commonest restrictive actions are to rogue out and burn infected plants and to destroy insects which transmit various viruses. Control of insects is also very effective in restricting some animal viruses. For example, elimination or reduction of the mosquito population is very helpful in controlling yellow fever in the human population and equine encephalitis in horses (vaccines against these viruses are also available). The destruction of diseased individuals has also been used to limit the spread of animal viruses, just as with plant viruses. An instance of this is the slaughter and burning of cattle infected with foot-and-mouth disease virus. In the human population, vaccines are currently the most practicable means of trying to control human virus diseases.

References

BOOKS

Barry, R. D., and B. W. J. Mahy (eds.): The Biology of Large RNA Viruses, Academic, New York, 1970.

Basic Mechanisms in Animal Virus Biology, Cold Spring Harbor Symp. Quant. Biol., vol. 27, New York, 1962.
Betts, A. O., and C. J. York (eds.): *Viral and Rickettsial Infections of Animals,* Academic, New York, 1967.
Cohen, S. S.: *Virus-induced Enzymes,* Columbia, New York, 1968.
Fenner, F.: *The Biology of Animal Viruses,* vols. 1 and 2, Academic, New York, 1968.
——— and D. O. White: *Medical Virology,* Academic, New York, 1970.
Hayes, W.: *The Genetics of Bacteria and Their Viruses,* 2d ed., Wiley, New York, 1968.
Hershey, A. D. (ed.): *The Bacteriophage Lambda,* Cold Spring Harbor Laboratory, Cold Spring Harbor, N.Y., 1971.
Horsfall, F. L., and I. Tamm (eds.): *Viral and Rickettsial Infections of Man,* 4th ed., Lippincott, Philadelphia, 1965.
Joklik, W. K., and D. T. Smith (eds.): *Zinsser Microbiology,* 15th ed., Meredith Press, New York, 1972.
Kenney, F. T., B. A. Hamkalo, G. Farelukes, and J. T. August (eds.): *Gene Expression and Its Regulation,* Plenum, New York, 1973.
Levy, H. B. (ed.): *The Biochemistry of Viruses,* Marcel Dekker, Inc., New York, 1969.
Mathews, C. K.: *Bacteriophage Biochemistry,* Van Nostrand Reinhold, New York, 1971.
Matthews, R. E. F.: *Plant Virology,* Academic, New York, 1970.
Mechanism of Protein Synthesis, The, *Cold Spring Harbor Symp. Quant. Biol.,* **34**, Cold Spring Harbor, N.Y., 1969.
Rothfield, L. I. (ed.): *Structure and Function of Biological Membranes,* Academic, New York, 1971.
Stent, G. S.: *Molecular Biology of Bacterial Viruses,* Freeman, San Francisco, 1963.
———: *Molecular Genetics,* Freeman, San Francisco, 1971.
Watson, J. D.: *Molecular Biology of the Gene,* 2d ed., W. A. Benjamin, Inc., New York, 1970.
Wolstenholme, G. E. W., and M. O'Connor (eds.): *Strategy of the Viral Genome,* Churchill & Livingstone, London, 1971.

JOURNAL ARTICLES AND REVIEW PAPERS

Assembly

Eiserling, F. A., and R. C. Dickson: Assembly of Viruses, *Ann. Rev. Biochem.* **41**:796–502, 1972.
Fraenkel-Conrat, H.: Reconstitution of Viruses, *Ann. Rev. Microbiol.* **24**: 463–478, 1970.
Hohn, T., and B. Hohn: Structure and Assembly of Simple RNA Bacteriophages, *Adv. Virus Res.* **16**:43–98, 1970.
Kushner, D. J.: Self-assembly of Biological Structures, *Bacteriol. Rev.* **33**: 302–345, 1969.

Leberman, R.: The Disaggregation and Assembly of Simple Viruses, *Symp. Soc. Gen. Microbiol.* **18**:183–205, 1968.

Phillips, B. A.: In vitro Assembly of Poliovirus. II. Evidence for the Self-assembly of 14S Particles into Empty Capsids, *Virology* **44**:307–316, 1971.

Wood, W. B., R. S. Edgar, J. King, I. Lielausis, and M. Henninger: Bacteriophage Assembly, *Fed. Proc.* **27**:1160–1166, 1968.

Attachment

Allison, A. C., and R. C. Valentine: Virus Particle Adsorption. III. Adsorption of Viruses by Cell Monolayers and Effects of Some Variables on Adsorption, *Biochim. Biophys. Acta.* **40**:400–410, 1960.

Cords, C. E., and J. J. Holland: Alteration of the Species and Tissue Specificity of Poliovirus by Enclosure of Its RNA within the Protein Capsid of Coxsackie B1 Virus, *Virology* **24**:492–495, 1964.

Dahl, D., and C. A. Knight: Some Nitrous Acid–induced Mutants of Tomato Atypical Mosaic Virus, *Virology* **21**:580–586, 1963.

Hiebert, E., J. B. Bancroft, and C. E. Bracker: The Assembly in vitro of Some Small Spherical Viruses, Hybrid Viruses, and Other Nucleoproteins, *Virology* **34**:492–508, 1968.

Philipson, L.: The Early Interaction of Animal Viruses and Cells, *Prog. Med. Virol.* **5**:43–78, 1963.

Tamm, I.: The Replication of Viruses, *Am. Sci.* **56**:189–206, 1968.

Tolmach, L. J.: Attachment and Penetration of Cells by Viruses, *Adv. Virus Res.* **4**:63–110, 1957.

Valentine, R. C., R. Ward, and M. Strand: The Replication-cycle of RNA Bacteriophages, *Adv. Virus Res.* **15**:1–59, 1969.

Control of Virus Infections

See Books: Fenner; Joklik and Smith

Gordon, P., and E. R. Brown: The Antiviral Activity of Isoprinosine, *Can. J. Microbiol.* **18**:1463–1470, 1972.

Hermann, E. C., Jr., and W. R. Stinebring (eds.): Second Conference on Antiviral Substances, *Ann. N.Y. Acad. Sci.* **173**:1–844, 1970.

Kleinschmidt, W. J.: Biochemistry of Interferon and Its Inducers, *Ann. Rev. Biochem.* **41**:517–542, 1972.

Maugh, T. H., II: Influenza: The Last of the Great Plagues, *Science* **180**: 1042–1044, 1159–1161, 1215, 1973.

Penetration and Multiplication

Baltimore, D.: Polio Is Not Dead, *Perspect. Virol.* **7**:1–12, 1971.

Becker, A., and J. Hurwitz: Current Thoughts on the Replication of DNA, *Prog. Nucleic Acid Res. Mol. Biol.* **11**:423–459, 1971.

Bishop, J. M., and L. Levintow: Replicative Forms of Viral RNA. Structure and Function, *Prog. Med. Virol.* **13**:1–82, 1971.

Cairns, J.: DNA Synthesis, *Harvey Lect.*, ser. **66**:1–18, 1972.

Dunnebacke, T. H., J. D. Levinthal, and R. C. Williams: Entry and Release of Poliovirus as Observed by Electron Microscopy of Cultured Cells, *J. Virol.* **4**:505–513, 1969.

Otsuki, Y., I. Takebe, Y. Honda, and C. Matsui: Ultrastructure of Infection of Tobacco Mesophyll Protoplasts by Tobacco Mosaic Virus, *Virology* **49**:188–194, 1972.

Ross, A. F.: Systemic Effects of Local Lesion Formation, in *Viruses of Plants*, A. B. R. Beemster and J. Dykstra (eds.), pp. 127–150, North-Holland Publishing Company, Amsterdam, and Wiley, New York, 1966.

Schonberg, M., S. C. Silverstein, D. H. Levin, and G. Acs: Asynchronous Synthesis of the Complementary Strands of the Reovirus Genome, *Proc. Natl. Acad. Sci., U.S.A.* **68**:505–508, 1971.

Shatkin, A. J.: Viruses with Segmented Ribonucleic Acid Genomes: Multiplication of Influenza versus Reovirus, *Bacteriol. Rev.* **35**:250–266, 1971.

Sinsheimer, R. L.: Bacteriophage φX174 and Related Viruses, *Prog. Nucleic Acid Res. Mol. Biol.* **8**:115–170, 1968.

Spiegelman, S., N. R. Pace, D. R. Mills, R. Levisohn, T. S. Eikhom, M. M. Taylor, R. L. Peterson, and D. H. L. Bishop: The Mechanism of RNA Replication, *Cold Spring Harbor Symp. Quant. Biol.* **33**:101–124, 1968.

Sugiyama, T., B. D. Korant, and K. K. Lonberg-Holm: RNA Virus Gene Expression and Its Control, *Ann. Rev. Microbiol.* **26**:467–502, 1972.

Temin, H. M.: RNA-Directed DNA Synthesis, *Sci. Am.* **226**:24–33, 1972.

Weissmann, C., G. Feix, and H. Slor: In vitro Synthesis of Phage RNA: The Nature of the Intermediates, *Cold Spring Harbor Symp. Quant. Biol.* **33**:83–100, 1968.

Westphal, H., and R. Dulbecco: Viral RNA in Polyoma- and SV40-transformed Cell Lines, *Proc. Natl. Acad. Sci., U.S.A.* **59**:1158–1165, 1968.

Release

Dunnebacke, T. H., et al.: See Penetration reference.

Oshiro, L. S., H. M. Rose, C. Morgan, and K. C. Hsu: Electron Microscope Study of the Development of Simian Virus 40 by Use of Ferritin-labeled Antibodies, *J. Virol.* **1**:384–399, 1967.

Zee, Y. C., L. T. Talens, and A. J. Hackett: Localization of a Small Ribonucleic Acid Virus within Cytoplasmic Cisternae, *J. Virol.* **1**:1271–1273, 1967.

Transcription

Calendar, R.: The Regulation of Phage Development, *Ann. Rev. Microbiol.* **24**:241–296, 1970.

Chamberlin, M.: Transcription 1970: A Summary, *Cold Spring Harbor Symp. Quant. Biol.* **35**:851–873, 1970.

Transduction

See Books: Hayes, pp. 620–649; Stent, pp. 415–428.

Translation

McAuslan, B. R.: Enzymes Specified by DNA-containing Animal Viruses, in *Strategy of the Viral Genome,* G. E. W. Wolstenholme and M. O'Connor (eds.), pp. 25–38, Churchill & Livingstone, London, 1971.

Ptashne, M.: Phage Repressors, in *Strategy of the Viral Genome,* G. E. W. Wolstenholme and M. O'Connor (eds.), pp. 141–150, Churchill & Livingstone, London, 1971.

Viral Pathogenesis

See Books: Cohen; Joklik and Smith; Matthews; and Stent.

Levine, A. J., and H. S. Ginsberg: Mechanism by Which Fiber Antigen Inhibits Multiplication of Type 5 Adenovirus, *J. Virol.* **1**:747–757, 1967.

Milne, R. G.: Plant Viruses inside Cells, *Sci. Prog.* (Oxford) **55**:203–222, 1967.

Smith, H.: Mechanisms of Virus Pathogenicity, *Bacteriol. Rev.* **36**:291–310, 1972.

CHAPTER 5 COMPLEX INFECTION

In the previous chapter the simplest view of the process of viral infection was presented. In that view, the process of infection consists of the invasion of a cell by a foreign genetic system. The effect of this invasion is to modify more or less the functioning of the cell, with great or small consequences to its health but with the end result that more virus just like that which initiated the infection is produced. This is too restricted a view of viral potentiality. Various ingenious experiments have yielded results which show that the consequences of infection can be much more diverse and intricate than this simple formula suggests. For example, researchers have explored the consequences of simultaneous infection with two or more viral mutants (also called *strains* or *variants*), mixed infection with unrelated viruses, and mixed infection with infectious and uninfectious particles. Among the phenomena recognized in these studies, first with phages and later with other types of viruses, are mating, marker rescue (cross reactivation), and multiplicity reactivation. All these activities probably depend on the same process, namely, genetic recombination. In addition, the phenomena of complementation, phenotypic mixing, and interference have been observed in mixed infections.

5-1 Mating

MATING IS THE INTIMATE ASSOCIATION OF THE UNDAMAGED NUCLEIC ACIDS OF VIRAL STRAINS WHICH RESULTS IN EXCHANGE OF GENETIC MATERIAL, I.E., GENETIC RECOMBINATION

Mating and genetic recombination occur in mixed infection of *Escherichia coli* by morphologically and serologically similar bacterial viruses. Such recombinations, first established by Hershey and Rotman in the late 1940s, were the earliest observed among viruses. Genetic recombination has since been demonstrated with animal viruses and possibly with plant viruses.

In order to detect recombination it is necessary for each of the two viruses involved to have at least one unique genetic trait. In *E. coli* phages, such traits as plaque type and host range illustrate markers that have been useful in recombination studies (*r*, denoting rapid lysis of infected bacteria to give clear plaques is one symbol that has been used to indicate the character, plaque type; *h* has been used as a symbol to indicate ability to infect a particular host bacterium). It is customary to refer to the character of the wild type phage, or of any other phage with an identical trait, by addition to the marker symbol of a plus sign. Thus in the case of common, i.e., wild type, coliphage T2, the description with respect to plaque type and host range is $T2r^+h^+$. A mutant of T2 giving a different plaque type but infecting the same strain of bacteria is designated $T2rh^+$; a mutant giving the same type of plaque but infecting a different strain of *E. coli* is indicated by $T2r^+h$.

T2 phage multiplies in a strain of *E. coli* called B, but there are mutants of *E. coli* B which will not allow multiplication of T2. Such a bacterial strain is designated B/2 (B bar 2), meaning that this strain bars the multiplication of phage T2.

A case of mating may now be considered in which a mixed infection of *E. coli* B is made with two mutants of T2 phage, both of which can multiply in *E. coli* B but only one of which can replicate in B/2. The various products of the mixed infection appearing in the lysates of infected bacteria can be characterized by plating some of the lysate on *E. coli* B and B/2 separately and on a mixture of B and B/2. By noting the strains of *E. coli* infected and the nature of the plaques observed, it can be concluded that four different types of T2 phages

TABLE 5-1 **Products of Mixed Infection of *E. coli* Bacteria with Two Strains of Phage T2: T2r$^+$h and T2rh$^+$**

Plaque type observed on E. coli *strains*			
B	*B/2*	*B and B/2*	*Designation of product*
Small, clear	No plaques	Small, turbid	T2r$^+$h$^+$ (wild type)
Small, clear	Small, clear	Small, clear	T2r$^+$h
Large, clear	No plaques	Large, turbid	T2rh$^+$
Large, clear	Large, clear	Large, clear	T2rh

issued from the mixed infection, as summarized in Table 5-1. (Note especially the results of the plating on a mixture of *E. coli* B and B/2.) The production of different phages by this mixed infection can be summarized as follows: T2r$^+$h × T2rh$^+$ → T2r$^+$h, T2rh$^+$, T2r$^+$h$^+$, T2rh. From inspection of the products, it is clear that the two input parental phages were reproduced, but in addition there seems to have been mating between the parental phages with the resultant production of two other types of phages, one of which appears to be wild type (T2r$^+$h$^+$) and the other a double mutant with respect to wild type (T2rh). These latter two phages are called genetic recombinants, and their distinctive traits are perpetuated in successive cycles of multiplication when one or another of them is used as a source of infection.

How can one conclude that the novel products of mixed infection are recombinants rather than mutants of one or the other of the input phages? First, mutants display random alterations in one or more of numerous traits, whereas the products of mating show precisely the traits of the parental phages and in a reciprocal manner. Next, the quantitative aspects of mutation and recombination are distinctive. Phage mutants may appear spontaneously among progeny phages with a frequency of about 1 in 1,000 to 1 in 1 billion, whereas recombinants of the sort discussed above occur with a frequency of about 1 in 5, the parental types comprising the rest of the population and appearing in approximately the same proportions as in the initial mixture with which infection was started.

Recombination can be and probably often is more complex than the example given in Table 5-1. For instance, rather than two-factor crosses (crosses in which two traits are followed), multiple-factor crosses may be involved if the mutants used in mixed infection differ in several traits. Furthermore, even in a two-factor cross, it cannot be assumed that only one mating event occurs. In fact, in the T-even

(T2, T4, T6) coliphages it is quite likely that a series of recombinational events occurs, perhaps as many as five. The number of rounds of mating before the process is interrupted by cell lysis varies with different phages. It appears to be nearer to one with phage lambda.

If more than two mutants simultaneously infect the same cell, matings can occur among all of them, further enriching the yield of recombinational products. That such rounds of mating do occur is demonstrable, for example, in a triparental cross in which bacteria are infected simultaneously with three phages having the genetic characters ab^+c^+, a^+bc^+, and a^+b^+c, respectively. Among the phages issuing from this mixed infection, one type having a character from each of the three parents (*abc*) is observed.

The events described above emphasize the importance to these phages of the latent period in which viral nucleic acid accumulates in cellular pools where mating and genetic recombination can occur.

Mating between strains of poliovirus has been observed by Hirst and colleagues and by Cooper. Some data obtained by Hirst in crosses between serum inhibitor-resistant mutants of poliovirus which illustrate recombination with this virus are given in Table 5-2.

TABLE 5-2 **Yield of Recombinants from Mating of Two Serum Inhibitor-Resistant Mutants of Poliovirus**

*HeLa cells infected with**	*No. of large plaques and total plaques in presence of both ho and bo sera*	*Double mutants (ho-bo), %*
Ho	0/10,276	<0.01
Bo	2/4,280	0.047
Ho and Bo	9/2,460	0.37

* Ho designates the mutant producing large plaques on HeLa cells either in the presence or in the absence of inhibitory horse serum (wild type gives large plaques in the absence of serum and very small plaques in the presence of serum); bo is the comparable mutant giving large plaques in the presence or absence of inhibitory bovine serum. In the presence of both horse and bovine serum inhibitors, large plaques are expected if the double mutant, ho-bo, is present or if there is a mutation of ho or bo to ho-bo. No mutants of ho to ho-bo were found among 10,276 plaques, and only two mutants of bo to ho-bo in 4,280 plaques. Hence, the fact that the yield of ho-bo in mixed infection was approximately fifteen times the background yield of ho-bo from mutation of bo supports the conclusion that recombination occurred (it should be noted that plating the progeny resulting from mixed infection separately under each inhibiting serum showed that the ratio of parental types was about 1).

SOURCE: From Hirst, 1962.

The results of mating experiments with strains of plant viruses have been generally inconclusive with a very few exceptions. Some possible reasons for this include the following: (1) there is a tendency for strains of plant viruses (in contrast with phages and animal viruses) to interfere mutually with the reproductive process in mixed or superinfections; (2) plant virus nucleic acids are generally small, which means there are few genes, and this in turn limits the probability of recombinant formation; (3) there is a rather high rate of spontaneous mutation, which produces mutants that are difficult to distinguish from recombinants. However, Best's work on one of the larger plant viruses, tomato spotted wilt virus, indicates that mixed infection between two strains of this virus may have involved mating and the production of recombinants. An example of his results is given in Table 5-3. Three new strains, R_1, R_2, and R_3, were found as products of mixed infection with strains A and E. These strains, as indicated in the table, showed symptoms different from those of either A or E (or of any of the spontaneous mutants of A or E ever observed), but they combined some of the symptom characters of both parental strains. The recombinant strains also bred true in successive transfers, as would be expected.

TABLE 5-3 **Effect of Mixed Infection with Strains A and E of Tomato Spotted Wilt Virus on Some Genetic Characters of the Progeny**

*Strain of virus and phenotypic formula**	*Symptoms† on:*						
	Tomato				*Tobacco*		
	(Lycopersicon esculentum *Mill. var. Dwarf Champion*)				(Nicotiana glutinosa *L.*)		(Nicotiana tabacum *L. var. Blue Pryor*)
	A	*B*	*F*	*G*	*C*	*D*	*E*
A:ABcdefg	+	+	−	−	−	−	−
E:abCDEfg	−	−	−	−	+	+	+
R_1:aBCdefg	−	+	−	−	+	−	−
R_2:abCdEfG	−	−	−	+	+	−	+
R_3:abCdeFg	−	−	+	−	+	−	−

* A capital letter denotes the presence of a particular character and a small letter its absence.

† Symptoms observed on the three types of test plant are given the following letter designations: A, apical necrosis, B, leaf necrosis and pigmentation; F, nonnecrotic yellow disks; G, etch; C, systemic infection; D, nonnecrotic yellow disks in *N. glutinosa*; E, rosetting. R_1, R_2, and R_3 represent products of the cross, A × E.

The frequency of recombination of genes is, on the basis of probability considerations, proportional to the distance separating them, which in turn is dependent on the size and total number of genes. Consequently, it would be anticipated that the frequency of recombination would be low in a virus possessing only a few genes and relatively high for a virus possessing a large number of genes (collectively called a *genome*). In order to provide a basis for predicting relative frequencies of recombinational events and numbers of gene products to expect, a table has been prepared (Table 5-4) in which an estimate of the number of genes for several viruses has been made. The figures in the table must be considered only crude approximations, serving as a guide until more exact values can be determined. Nevertheless, the number of genes estimated for the various viruses can be compared with estimates for bacteria of a few thousand and of plant and animal cells of a few hundred thousand per chromosome.

5-2 *Marker Rescue (Cross Reactivation)*

MIXED INFECTION WITH A FULLY ACTIVE MUTANT AND A GENETICALLY DIFFERENT INACTIVATED MUTANT CAN RESULT IN GENETIC RECOMBINATION BETWEEN THE MUTANT NUCLEIC ACIDS. IN THIS MANNER GENETIC MARKERS FROM THE INACTIVATED VIRUS CAN APPEAR IN ACTIVE NUCLEIC ACID, I.E., CAN BE RESCUED

There is another type of mixed infection which differs from the mating reaction just considered in that it involves simultaneous infection, not with two fully infectious mutants but with one infectious mutant and a variably inactivated second mutant. The results of such a mixed-infection experiment performed with two mutants of coliphage T4 are represented in Fig. 5-1. It can be seen that ultraviolet light rapidly destroys the infectivity, as judged by the number of wild type (r^+) plaques obtained when the T4r^+ phage is irradiated and then tested alone in *E. coli* cells. However, the loss of infectivity with the same dose of ultraviolet light is greatly reduced when a mixture of ultraviolet-treated and untreated mutant, T4r, is used to infect cells in the plaque assay. The curve in Fig. 5-1 labeled "r^+ and r" shows only the number of r^+ plaques (not r^+ and r plaques) obtained in the mixed infection made after the T4r^+ had been irradiated. Since the

TABLE 5-4 **Estimate of Number of Genes in Some Viral Nucleic Acids***

Virus	*Estimated no. of genes*
Tobacco necrosis satellite	1[†]
Brome mosaic	3
Coliphage f1	3
Coliphage ϕX174	7
Polyoma	7
Potato virus X	7
Tobacco mosaic	7
Tobacco necrosis	7
Turnip yellow mosaic	7
Shope papilloma	8
Simian virus 40	8
Influenza	10
Poliovirus	10
Reovirus	17
Cytoplasmic polyhedrosis	22
Rous sarcoma	33
Adenovirus	38
Salmonella P22	47
Coliphage lambda	53
Herpes simplex	128
Coliphage T4	200
Tipula iridescent	260
Vaccinia	267

* Values were obtained by dividing the molecular weight of the nucleic acid as given in Table 3.3 by 0.3×10^6. If the nucleic acid is double-stranded, the value was further divided by 2 on the assumption that only one of the two strands is transcribed. The basic assumptions reflected in the factors used to make this very approximate calculation of gene number are that all of the coding strand of the nucleic acid is transcribed, that the average weight of a nucleotide is 300, that the code is a nonoverlapping triplet code (i.e., three nucleotides code for one amino acid), that the proteins coded for have a molecular weight of 35,000, and that the average amino acid has a molecular weight of 100.

† Since it is known that the satellite protein is only about half the size of the assumed figure of 35,000, it appears that satellite RNA could have two genes instead of one, the alternative being that the amount of RNA in excess of that needed for coat protein does not code for anything.

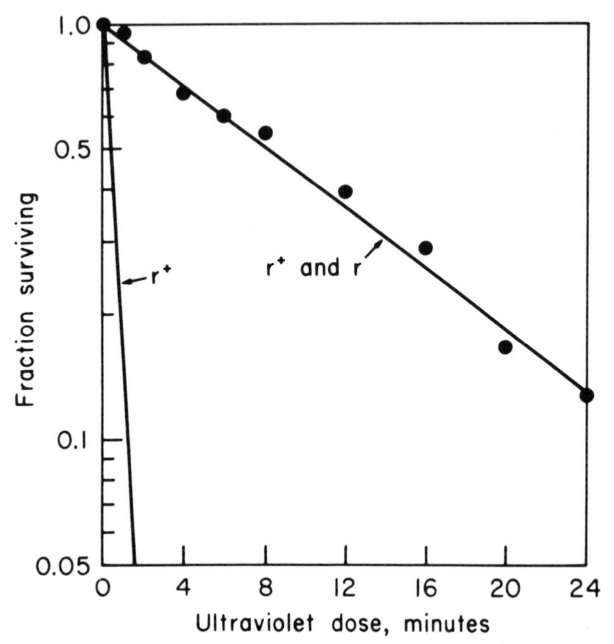

FIG. 5-1. **Rescue of r$^+$ marker by mixed infection of** *E. coli* **with ultraviolet-irradiated T4r$^+$ and unirradiated T4r. (Adapted from Doermann et al., J. Cell Comp. Physiol. 45, 51–74, 1955.)**

number of r$^+$ plaques is clearly much larger in the mixed infection for a given dose of ultraviolet light than in the test of ultraviolet-irradiated T4r$^+$ alone, the term *marker rescue* was applied, the marker rescued being the wild type plaque, r$^+$, in this case. The other term sometimes used, *cross reactivation,* reflects that there seems to be a reactivation of r$^+$ marker as a consequence of crossing, i.e., mixed infection with the irradiated and unirradiated phages.

Cross reactivation has also been observed with animal viruses. For example, Kilbourne and associates infected chick embryos with active A_2 influenza virus and heat-inactivated A_0 influenza virus. Obtained among the progeny was a viral strain having the hemagglutinin and internal antigen of the A_0 parent and the neuraminidase of the A_2 strain. Thus, the internal antigen and the hemagglutinin of the A_0 strain were rescued. Similarly, marker rescue has been reported as a result of infection of mouse cells with a mixture of active and ultraviolet-inactivated rabbitpoxvirus.

That cross reactivation involves a series of mating events, as it does when both mutants are fully active, is indicated by the finding of Doermann and collaborators that mixed infection with phage mutants differing in several characteristics gives recombinants differing in the markers that have been "rescued" from the irradiated mutant.

5-3 Multiplicity Reactivation

IF A VIRUS HAS BEEN INACTIVATED BY THE PRODUCTION OF RANDOM LESIONS IN ITS NUCLEIC ACID, INFECTION WITH MANY SUCH VIRUS PARTICLES CAN RESULT IN GENETIC RECOMBINATION BETWEEN DEFECTIVE NUCLEIC ACIDS TO PRODUCE FULLY ACTIVE GENOMES

A third type of recombinational mixed infection has been observed in which two of the same kind of ultraviolet-inactivated viruses, neither of which is infectious by itself, can produce an infection if both get into the same cell. This phenomenon, first observed by Luria and Dulbecco with T2 coliphage, is called *multiplicity reactivation.* It has also been observed with several animal viruses such as influenza, fowl plague, and Newcastle disease viruses as well as with reovirus and certain poxviruses. In most cases the viruses tested were inactivated with ultraviolet light; however, multiplicity reactivation was also shown by Schäfer and Rott to occur with fowl plague virus after inactivation with hydroxylamine.

Mating between viral strains, cross reactivation (marker rescue), and multiplicity reactivation are all considered to take place by the same means, namely, genetic recombination. The process in each case can be visualized as shown in Fig. 5-2. However, the sketches in Fig. 5-2 merely summarize some major stages of genetic recombination without attempting to indicate the mechanisms involved. The detailed molecular mechanisms are yet to be resolved; some of the complexities of the recombinational process have been considered in review articles (e.g., see Clark, and Davern). At this stage, perhaps the only certain molecular feature of recombination in viral systems is that strand breaks occur in double-stranded viral nucleic acids, followed by partial degradation and then repair. If breakage and reunion occur within the same strands or between comparable segments of identical genomes, the recombinational process can go undetected. When genomes which are very similar but differ in a few distinctive markers (at least two) undergo breakage when in close

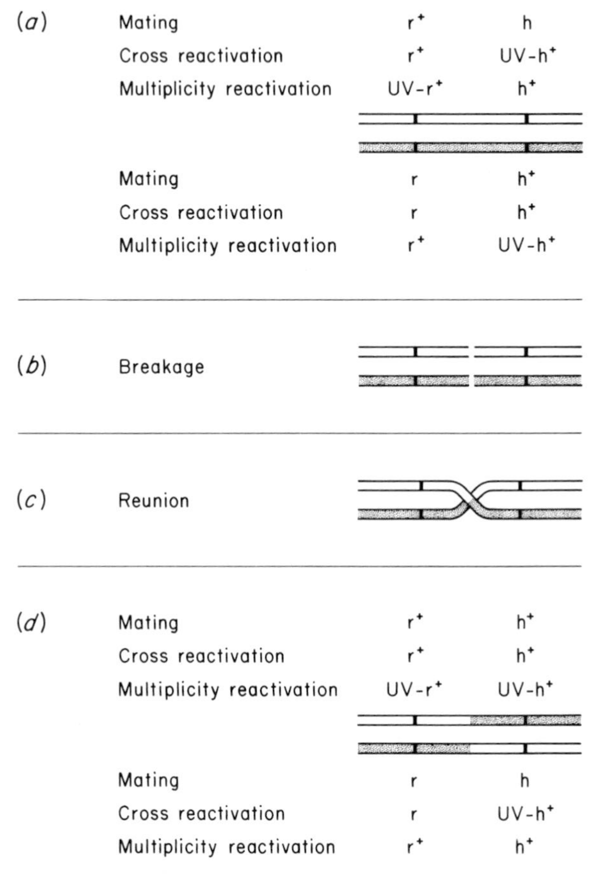

FIG. 5-2. **Three types of interaction that occur with a pair of phage genomes, all of which involve genetic recombination: (a) the initial pair of genomes; (b) breakage; (c) crossing-over; (d) the recombinants. The pertinent genetic loci in the three examples of genetic recombination are listed above and below the regions of the genomes in which they occur.**

proximity, rejoining can result in an exchange of detectably different pieces of genome. All this involves a series of enzyme-mediated events. Thus there are endo- and exonucleases which cut nucleic acid strands and degrade lengths of exposed single-stranded pieces, and there are other enzymes which synthesize new segments and restore structures to double-stranded form. Finally there are ligases which rejoin ends. At least some of these enzymes appear to be coded for by the viral genome.

5-4 Phenotypic Mixing

WHEN DIFFERENT VIRUSES MULTIPLY IN THE SAME CELL, THEIR PRECURSOR PARTS ARE MIXED IN THE METABOLIC POOLS AND SOME PARTICLES MAY BE ASSEMBLED FROM COMPONENTS OF DIFFERENT VIRUSES

In contrast to the heritable changes resulting from recombination of nucleic acids in mixed infection, it is also possible to obtain phenotypic changes that are transient since they do not involve the viral genomes. This phenomenon, called *phenotypic mixing,* can be illustrated by the results of mixedly infecting strains of *E. coli* B with phages T2 and T4. Most of the progeny issuing from joint infection of *E. coli* B will infect either *E. coli* B/2 or B/4 as well as the original *E. coli* B. Some of these progeny are, of course, the parental types, but others are not. For example, some of the phages which infect B/4 are parental type T2, but some of them, upon further passage, produce pure T4. The explanation of the latter type is that in mixed infection, parts of T2 and T4 occur together in intracellular pools, and since these phages are very similar in size and structure, some parts get interchanged in the process of assembling complete phage particles. The crucial exchange in the above example is in tail parts, since these determine the strains of *E. coli* cells to which such a phage can attach and consequently can infect. Hence if particles containing T4 DNA acquire T2 tail fibers, they will be able to infect B/4 cells. However, only T4 parts will be produced in this next cycle of replication, and hence the progeny will have the usual tail parts, host range, and other properties of T4.

A more complicated form of phenotypic mixing in the T2-T4 system was observed by Streisinger. Some of the progeny appeared to differ in host range from either parental type in that for one cycle of infection they infected both B/2 and B/4 cells. Accordingly such viruses must possess tail parts characteristic of both T2 and T4.

Such nonheritable, doubly specific surface structures are quite commonly observed among the products of mixed infection with animal viruses, including myxo- and paramyxoviruses (influenza, Newcastle disease, measles, and Sendai) and neurotropic viruses such as Sindbis and poliovirus. However, the initially observed type of phenotypic mixing in which the infectivity-determining external features are solely those of one virus has also been observed in mixed infections with poliovirus and Coxsackie virus, and in mixed infection with mutants of herpes simplex virus.

A demonstration of phenotypic mixing of plant viruses was shown by Rochow with two serologically unrelated isolates of barley yellow dwarf virus. This example is of additional interest because it also bears on the specificity of transmission of viruses by insect vectors. Each of the isolates of yellow dwarf virus is transmitted specifically by a different aphid, but if one of the aphids is fed on leaves or extracts of doubly infected plants, it transmits both viruses, rather than just one of them. Serologic testing combined with feeding experiments indicates that in a mixed infection some of the RNA of both viruses gets coated with the protein of one of them. This apparently makes it possible for the aphids to transmit both viruses.

In another set of examples of phenotypic mixing, observed by Kassanis and Bastow, the RNAs of strains of tobacco mosaic virus defective in ability to produce a useful coat protein were coated in vivo in a mixed infection with the protein from another strain of TMV. The borrowed coat protected the RNA of the defective viruses, and hence their infectivity was greatly enhanced.

5-5 *Complementation*

RELATED OR UNRELATED, DEFECTIVE VIRUSES CAN MULTIPLY IN A CELL IF THE COMBINATION OF PARTICLES MAKES UP FOR DEFICIENCIES; SOME VIRUSES REQUIRE TWO OR MORE FUNCTIONALLY DIFFERENT PARTICLES, AKIN TO CHROMOSOMES, IN ORDER TO MULTIPLY

In studies on coliphage T4, Benzer found that certain spontaneous mutants could individually multiply in *E. coli* strain B but not in *E. coli* strain K12. However, in mixed infections of bacterial strain K12 with two mutants such as rIIA and rIIB, in which the mutations involved different functions, both parental types multiplied (also producing some recombinants). Apparently each mutant complemented the other by supplying a missing function; consequently both mutants could multiply in a host in which neither could grow by itself. If the mutations involved the same function and the same gene, as with strains rIIA1 and rIIA2, then there was no complementation, and multiplication in K12 bacteria failed to occur.

Complementation has since been observed extensively in bacterial virus systems and in fact has become a useful tool in gene mapping (see Streisinger et al., 1964). Mutations in the same gene do not complement each other, whereas mutations in separate genes do. This is particularly helpful in distinguishing between several genes which may be involved in the same function.

Complementation appears to be fairly common among animal viruses, although in some cases it is better known by other terms such as defective and helper viruses, satellites, transcapsidation, etc. The kind of complementation shown to occur between defective mutants of bacterial viruses also occurs between defective mutants of several animal viruses. A common type of defective mutant of animal viruses, as well as bacterial viruses, is the temperature-sensitive mutant (*ts* mutant). These *ts* mutants do not multiply or multiply only poorly at temperatures at which the wild type strain will replicate fairly well. Presumably some gene product or function is deficient in the *ts* mutant at the restrictive temperature. However, mixed infection with *ts* mutants in which mutations are in different genes results in multiplication of both mutants, each mutant supplying the function deficient in the other. A good illustration is the case of rabbitpoxvirus. Many *ts* mutants of this virus were obtained by Padgett and Tomkins by growing the virus in the presence of bromodeoxyuridine. Eighteen such mutants were selected which gave 10^{-5} the plaque counts on plates of pig kidney cells at the restrictive temperature (39.5°C) that were obtained at the permissive temperature (34.5°C). Mixed infections between pairs of the selected mutants gave the results shown in Table 5-5. As indicated by the results in the table, complementation was manifested in mixed infection in all but six pairs by a yield of progeny virus which was greatly increased over single infection (fivefold to 90-fold).

Another example of complementation in an animal virus system involves viruses of the avian leukosis group. The Bryan high-titer strain of Rous sarcoma virus (RSV) transforms chick embryo cells in tissue culture so that they overgrow the usual monolayer of cells to form little mounds (foci) of tumorous cells. These foci fail to produce infectious virus, although the cells continue to multiply and when transplanted into young chicks they cause the appearance of tumors in 5 days and death from malignant growth by 10 days. If another virus of the avian leukosis group, such as avian lymphomatosis virus or avian myeloblastosis virus, is added to RSV-transformed cells, then, as Rubin and the Hanafusas showed, large amounts of both RSV and the "helper virus" are produced (Fig. 5-3). It appears, therefore, that the Bryan strain of RSV is a defective virus

TABLE 5-5 **Complementation between *ts* Mutants of Rabbitpoxvirus***

	1	2	3	4	5	6	7	8	9	10	11	12	13	14	15	16	17	18
1	75																	
2	450	32																
3	300	840	53															
4	ND†	240	ND	110														
5	210	166	140	ND	37													
6	650	810	900	ND	156	10												
7	160	340	380	ND	240	430	45											
8	210	320	500	**52**	220	200	150	28										
9	610	520	610	**125**	212	920	160	**52**	60									
10	188	192	106	ND	ND	ND	ND	ND	ND	21								
11	260	670	290	230	ND	ND	ND	390	330	ND	11							
12	530	940	700	ND	170	210	270	92	640	59	77	2						
13	180	260	320	ND	190	290	290	130	158	154	180	172	57					
14	155	390	80	430	ND	ND	ND	320	400	ND	510	120	154	16				
15	**80**	510	340	ND	158	190	280	92	250	ND	ND	60	133	ND	2			
16	250	1040	163	176	ND	ND	199	340	ND	84	940	146	230	96	ND	0.4		
17	143	580	270	ND	ND	ND	178	126	ND	189	350	40	310	96	ND	200	1	
18	193	270	210	ND	250	130	190	**29**	**63**	ND	ND	198	160	ND	88	ND	ND	39

* 2×10^5 PK (pig kidney) cells were infected with an input multiplicity of 2.5 PFU (plaque-forming units) per cell of both the parental viruses under test; control tubes were infected with an input multiplicity of 5 PFU per cell of one parent only. After adsorption, residual inocula were removed and the cell monolayers were washed once and then incubated at 39.5°C for 20 hr. Yields were titrated at 34.5°C and in some cases at 39.5°C on PK cell monolayers. Results are shown as the mean value of the yield ($\times 10^4$ PFU) from three replicate experiments; controls are recorded on the diagonal of the table. Complementation was observed in most cases tested (1.4-fold greater than parental yield); the exceptions are boldface figures, which indicate the cases in which complementation was not observed.

† ND = not done.

SOURCE: Adapted from Padgett and Tomkins, *Virology* 36, 161–167, 1968.

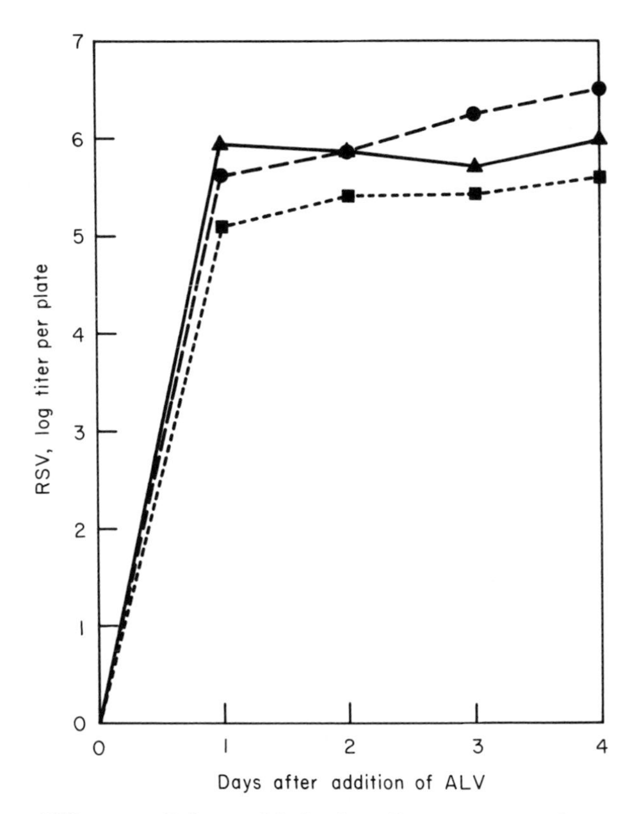

FIG. 5-3. **Release of infectious Rous sarcoma virus (RSV) from tissue cultures of chick embryo cells following addition of about 1,500 infectious units of an avian leukosis helper virus (ALV) per cell. Titer of the RSV released into the medium was assayed at daily intervals. The results obtained with three different cultures of chick cells are shown. (From Hanafusa et al., 1963.)**

in certain chick embryo cells but that other avian leukosis viruses complement RSV in mixed infection and supply the missing function. It is likely that a viral envelope protein which is essential in the attachment step of infection is missing in this kind of defective RSV particle. However, the broader relationship among avian tumor viruses and their host cells are much more intricate than the simple case cited above (see Vogt, 1971).

Mixed infection of rabbits with active Shope fibroma virus and heat-inactivated myxoma virus yields what for years was a puzzling result called the *Berry-Dedrick phenomenon* after its original observers. The phenomenon appears now to be another example of complementation between viruses. The fibroma virus, originally observed by Shope in wild cottontail rabbits of the Eastern United States, where it causes a benign skin tumor that regresses in 10 to 14 days, causes a similar mild disease in domestic rabbits. In contrast, the myxoma virus causes a highly virulent systemic disease in domestic rabbits that is about 98 percent fatal. When domestic rabbits are mixedly infected with live fibroma virus and heat-inactivated myxomatosis virus, some of the animals die with typical myxomatosis. This is the Berry-Dedrick phenomenon. The generally accepted explanation of this phenomenon is that the genome of the myxoma virus is not damaged by the heat treatment but that the mechanism for uncoating the genome intracellularly has been destroyed. Myxoma and fibroma are morphologically similar viruses of the poxvirus group; these viruses have inner and outer protein coats. The outer coat is readily removed in the infectious process by cellular enzymes, but the inner coat is removed by a rather specific virus-induced protease which arises as follows. Poxvirus particles contain an RNA polymerase which is able to transcribe a portion of the viral DNA before the viral genome is released from its inner protein coat. This transcription presumably results in the protease needed to remove inner coat from the viral genome. Returning to the Berry-Dedrick phenomenon, the myxoma RNA polymerase is denatured by heat treatment, but in mixed infection this enzyme is supplied for both viruses by the fibroma virus, thus releasing both myxoma and fibroma genomes in functional states. The severe consequences of the resulting myxomatosis completely obliterate the effect of the fibroma virus.

Another instance of complementation is one which may have implications for such areas of virology as cellular transformation and oncogenicity, host specificity, and formation of virus hybrids. This is the interaction of adenovirus and simian virus 40 (SV40). Both these viruses are potentially oncogenic, although they seldom induce tumors in their presumed natural hosts. For instance, SV40 has no known oncogenic action in monkeys but readily induces tumors in and/or transforms cells of a series of unnatural hosts such as mouse, rat, rabbit, hamster, mongoose, lemur, guinea pig, cat, dog, cow, pig, and man. Similarly, human adenoviruses that do not transform human cells do transform hamster cells. Mixed infections with SV40 and each of several adenoviruses raise the yield of infectious adenovirus to about a thousand times the amount produced in single infections (Table 5-6) and also result in the formation of mixed particles containing components of both viruses. Use of other viruses such as herpes, measles, Shope rabbit papilloma, and human wart viruses fails to enhance the replication of adenovirus type 7. Hence the requirement for SV40 appears to be specific for complementation of adenoviruses in green monkey kidney cells.

Conversely, adenovirus is needed for the reproduction of a kind of hybrid virus that emerges from mixed infection with adenovirus type 7 and SV40. This hybrid virus (called *PARA,* meaning Particle

TABLE 5-6 **Some Features of Single and Mixed Infection of Green Monkey Kidney Cells with Adenovirus and Simian Virus 40**

Virus inoculated	*Adenovirus infectious titer (log_{10} PFU*/ml)*	*Adenovirus-induced antigens*†		*SV40-induced antigens*	
		Tumor	*Viral*	*Tumor*	*Viral*
Adenovirus type 2	4.0	+	−	−	−
Adenovirus type 7	3.8	+	−	−	−
SV40		−	−	+	+
SV40 and adenovirus type 2	7.4	+	+	+	+
SV40 and adenovirus type 7	6.8	+	+	+	+

* PFU = plaque-forming units, tested at 72-hr postinoculation. The initial titers (1-hr postinoculation) for the adenoviruses in the absence of SV40 were 4.9 and 4.7 for types 2 and 7, respectively. Comparison of these figures with those in the table for 72 hr illustrated the failure of adenoviruses to multiply in green monkey kidney cells.

† The respective antigens were detected by the fluorescent antibody technique. A + score means that the antigen was present in readily detectable amounts; a minus designation means that the antigen was not present or present in concentrations too low to score as positive.

SOURCE: Adapted from Rapp and Melnick, 1966.

TABLE 5-7 **Some Properties of Adenovirus, PARA, and Simian Virus 40 (SV40)**

Property	*Adenovirus*	*SV40*	*PARA*	*PARA + adenovirus*
Replication in green monkey kidney cells	−	+	−	+
Replication in human embryonic kidney cells	+	±	−	+
Presence of adenovirus coat protein	+	−	+	+
Induction of adenovirus coat protein	+	−	+	+
Induction of SV40 coat protein	−	+	−	−
Induction of adenovirus tumor (T) antigen	+	−	−	+
Induction of SV40 tumor (T) antigen	−	+	±	+

SOURCE: Adapted from Rapp and Melnick, 1966.

Aiding Replication of Adenovirus) is defective in ability to reproduce and requires complementation from a nonhybridized adenovirus for its growth. In turn PARA, as its name indicates, enables adenoviruses to multiply in monkey kidney cells where normally they are poorly infectious. PARA appears to have a partial SV40 genome and an adenovirus protein coat. Any of a series of adenoviruses can complement PARA, but the progeny PARA particles then issue with a protein coat characteristic of the adenovirus type used in the complementation. For example, a PARA with adenovirus type 2 protein coat added to adenovirus type 7 in a mixed infection will result in the production of PARA with adenovirus type 7 protein coat. The term *transcapsidation* has been proposed for this phenomenon. A summary of some of the properties of PARA, adenovirus, and SV40 is presented in Table 5-7.

An instance of apparent complementation between strains of tobacco mosaic virus was used by Kado and Knight in locating the coat-protein gene of TMV. The half of TMV RNA terminating with a free ribosyl 5′-OH was suspected of being the locus for the coat-protein gene of TMV. Residual half-particles of TMV containing the 5′ portion of the RNA were obtained by selective stripping of protein subunits from the 3′ half of the TMV particles and clipping the exposed RNA with nuclease. When these were tested alone on tobacco plants, no sign of infectivity was obtained and no viral coat protein could be isolated. However, when plants were mixedly infected with half-particles of TMV and whole infectious particles of the HR strain of TMV, TMV coat protein, but no whole TMV, was found among the products of the mixed infection. The TMV coat protein was identified by its distinctive behavior in agar gel serology and electrophoresis in polyacrylamide. Thus the HR strain appears to have supplied some function needed for the production of TMV protein from the message in the provided half of the TMV RNA. At the same time the fact that TMV protein was produced confirms the location of the coat-protein gene in the 5′ half of the TMV RNA.

Plant viruses are widely viewed as the simplest viruses structurally and genetically. In many cases this concept can be supported. However, several plant viruses are now known (Table 5-8) whose genomes are distributed among two to four discrete particles, all of which are required for full functioning of the virus in question. In one sense this is an example of intraviral complementation. Alternatively, when compared with higher organisms on a functional basis, it appears that these viruses are comprised of two or more chromosomes, these chromosomes multiplying much more extensively than their cellular counterparts.

Tobacco rattle virus was probably the first well-documented instance of a multichromosomal plant virus. This virus has a wide host range in Europe and North America, including more than 100

TABLE 5-8 **Some Multiparticulate Plant Viruses**

Virus	*No. of discrete particles (chromosomes)*
Squash mosaic group:	
Squash mosaic	2
Bean pod mottle	2
Cowpea mosaic	2
Tomato top necrosis	2
Tobacco rattle	2
Tobacco streak	2
Brome mosaic	3
Alfalfa mosaic	4

species of crop plants, ornamentals, and weeds. Two peculiarities had long been associated with the virus. First there seemed to be "stable" ("multiplying") and "unstable" ("nonmultiplying" or "winter-type") forms of the virus. The symptoms appearing in a given host plant appear identical for the two forms, but the disease is readily transmitted from lesions containing the stable type of virus and very poorly from lesions representing the unstable type. The two types of virus also behave quite differently in systemic hosts, the one form being much more invasive and productive of virus than the other.

The second, long-known, curious feature of tobacco rattle virus is that the stable type is characterized by two classes of elongated particles. All the particles are about 25 nm in diameter, but their lengths are distributed about two modes, which, for example, are 66 and 183 nm for the PRN strain and 50 and 172 nm for the SAL strain. Since the lengths of the longer particles are not, for most strains, even multiples of the short ones, it seems unlikely that one is derived from the other by mechanical shearing; thus the enigma of two types of infection and two lengths of particles.

This puzzle was solved by several groups of workers (see reviews by Sänger and by Lister), all of whom showed that the unstable type of tobacco rattle virus consisted of viral RNA devoid of protein coat, while the stable form had such a protective coat. The relationship between the long and short particles of rattle virus was elucidated by determining the symptoms and products associated with isolated viral RNA, and by noting the results of mixed infection with long particles of one strain and short particles of another. Such a series of experiments is summarized in Fig. 5-4.

FIG. 5-4. **Products observed and symptoms produced by components and pairs of components of two strains of tobacco rattle virus: complete tobacco rattle virus, German strain (C-TRV-Ger); defective tobacco rattle virus, German strain (D-TRV-Ger); complete tobacco rattle virus, United States of America strain (C-TRV-USA); and defective tobacco rattle virus, United States of America strain (D-TRV-USA). The specific markers of the systems (particle size and shape, type of lesion on Nicotiana tabacum L. var. Xanthi-nc, and type of coat protein when present) are represented diagrammatically. The sedimentation profiles indicate the relative positions of various sedimenting species in a sucrose density gradient after 2 hr at 24,000 rpm. The complete viruses were purified from infected tobacco plants by means of differential centrifugation, while the defective viruses were extracted from infected tobacco with TEM buffer (tris:EDTA:mercaptoethanol) in the presence of bentonite and subsequently precipitated from the extract with ethanol. (Adapted from Sänger, 1968.)**

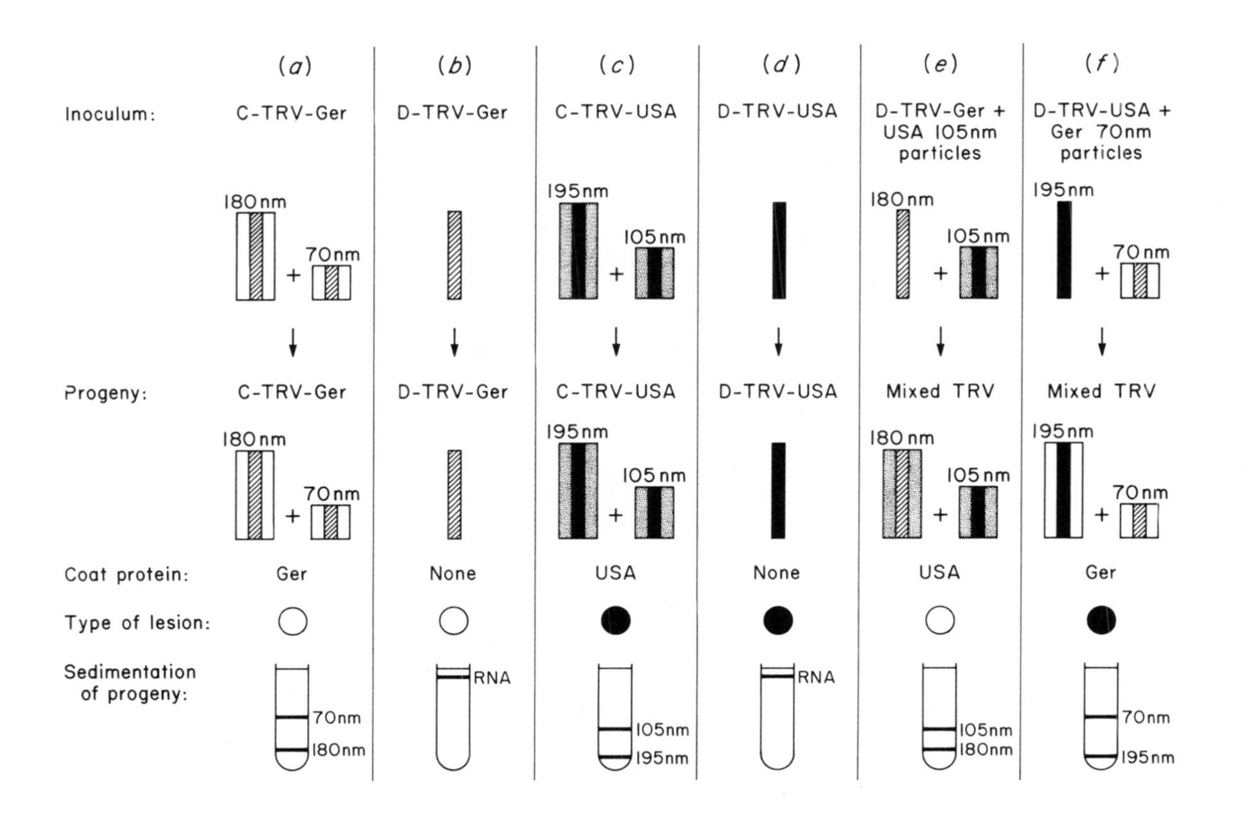

(a)
(b)
(c)
(d)
(e)
(f)
Inoculum:
C-TRV-Ger
D-TRV-Ger
C-TRV-USA
D-TRV-USA
D-TRV-Ger + USA 105nm particles
D-TRV-USA + Ger 70nm particles
180nm
70nm
195nm
105nm
Progeny:
C-TRV-Ger
D-TRV-Ger
C-TRV-USA
D-TRV-USA
Mixed TRV
Mixed TRV
Coat protein:
Ger
None
USA
None
USA
Ger
Type of lesion:
Sedimentation of progeny:
RNA
70nm
180nm
105nm
195nm

As indicated in the figure (columns *a* and *c*), RNA ensheathed in coat protein, in the form of particles of two predominant lengths, and called complete virus, was obtained with either the German or the United States strain only when the inoculum contained both long and short particles. When defective virus (RNA equivalent to that in long particles) (columns *b* and *d*) was used as the inoculum, characteristic symptoms and more viral RNA but no viral coat protein were produced in tobacco. Thus it appears that with each strain of rattle virus a homologous complementation occurs between short and long particles, the long particles providing the RNA-replicase function for both, while the short particles provide the coat-protein gene. This concept was strengthened by the outcome of heterologous complementation tests made by mixedly infecting tobacco with long particles of one strain and short particles of another. As summarized in Fig. 5-4, columns *e* and *f*, the symptoms produced in mixed infections are always characteristic of the RNA of the long particles, while the coat protein, as determined by serologic tests, is typical of the short particles.

The complementation observed with alfalfa (lucerne) mosaic is more complex than that of tobacco rattle virus. First, there are four classes of bacilliform particles differing specifically in length, and none of these particles is infectious by itself. Next, the pattern of complementation productive of infection is somewhat bizarre in that only three of the four nucleoprotein particles are required for infection, whereas when the isolated nucleic acids are employed for inoculum, all four classes of RNA are required. It has been postulated by Bol, Van Vloten-Doting, and Jaspars that the coat protein of this virus (which is identical in the four classes of particles) is the gene product of the RNA of the smallest nucleoprotein and that this is somehow crucial in the initiation of infection. Consequently infection can be started with three different mixtures: (1) the three largest nucleoprotein species, (2) the RNAs of the three largest nucleoprotein species plus added coat protein, (3) the four RNAs of the four different nucleoproteins. The precise nature of this complementation remains to be shown, but it appears that the critical function of the coat protein is not protection of the various RNAs from nucleases, since its infection-enhancing qualities are asserted at a level of only four to eight protein subunits per RNA molecule, far less than the number expected to provide protection from nucleases.

Another complex example of complementation between plant viruses was observed by Bancroft in joint infections with brome mosaic (BMV) and cowpea chlorotic mottle (CCMV) viruses. Though

BMV and CCMV share many physical properties (size, shape, multiparticulate nature, and RNA content), they appear to be serologically distinct and hence have quite different coat proteins. They also have dissimilar host ranges, so that one will infect many plants not susceptible to the other. However, both infect certain species of *Chenopodium* plants. From whatever susceptible plants employed, preparations of BMV or CCMV can be obtained which contain particles of uniform size but which have different species of RNA. These different species of RNA, named 1, 2, 3, and 4, have molecular weights of about 1.09, 0.99, 0.75, and 0.28×10^6 for BMV; and 1.15, 1.0, 0.85, and 0.32×10^6 for CCMV. Species 1, 2, and 3 are required to initiate BMV infections, and probably the same is true for CCMV, although some infection seems to occur with only species 1 and 2. Species 4 is not essential for infection in either case. *Chenopodium hybridum* inoculated with a mixture of BMV species 1, 2, and 3 RNA displays many large necrotic lesions, whereas when the inoculum is CCMV RNA species 1, 2, and 3, the lesions are somewhat smaller. When *C. hybridum* is inoculated with a mixture of BMV species 1 and 2 RNA and CCMV species 3 RNA, lesions considerably smaller than those invoked by either BMV or CCMV appear in numbers equivalent to about 10 percent of the value obtained for the homologous BMV series. These lesions contain hybrid virus, and this virus appears from the results of serologic tests to have the coat protein of CCMV. The providing of coat protein by the CCMV RNA can be considered complementation of the BMV RNA species present.

A curious instance of complementation has been observed in both plant and animal systems in which a virus enables a small, serologically unrelated particle to replicate in a given host which otherwise excludes it. The first pair of this sort to be recognized was tobacco necrosis virus and its satellite virus (see Kassanis, 1968).

The necrosis virus is a spheroidal, nucleoprotein particle about 28 nm in diameter which contains about 1.5×10^6 daltons of single-stranded RNA (representing about five genes); the satellite is a spheroidal, nucleoprotein particle which contains about 0.4×10^6 daltons of RNA (representing one or two genes). In nature, the two types of particle often occur together in infected plants, where the satellite particles may outnumber the virus particles; they also appear together in purified preparations of necrosis virus. However, the two types of particles can be separated, and it can be shown that the necrosis virus in some cases replicates indefinitely without reappearance of satellite. In contrast, there is no evidence that the satellite can multiply at all in the absence of necrosis virus.

Several strains of tobacco necrosis virus are known which are serologically related but distinguishable; likewise there are several strains of satellite which are serologically related but distinguishable. However, no satellite appears to be serologically related to any necrosis virus, indicating that the satellite RNA contains a gene for its own coat protein.

A strikingly parallel situation occurs with adenoviruses, where spheroidal satellites about 22 nm in diameter are found together with spheroidal adenovirus particles about 75 nm in diameter. These particles are called *adeno-satellite virus* or *adeno-associated virus*. The adeno-satellites differ from the necrosis satellites in that they contain double-stranded DNA, rather than single-stranded RNA, and relatively more of it, i.e., equivalent to about five genes. Adeno-satellites also appear less commonly associated with their helper virus in field isolates (in contrast to laboratory cultures) than the necrosis satellites. However, the following points of similarity exist: both satellites have particles smaller than those of their helper viruses and contain less nucleic acid, i.e., have fewer genes; both inhibit replication of their helper viruses; both occur in serologically specific types which have no serologic relationship to their helpers; and both can persist for some days in cells without replicating and then multiply upon addition of helper virus. Finally, necrosis and adeno-satellites resemble each other in that their origin and significance are obscure.

5-6 Interference

SOMETIMES A VIRUS WILL INTERFERE WITH THE MULTIPLICATION OF ANOTHER VIRUS

As noted in the consideration of mating and other instances of genetic recombination, closely related coliphages can multiply in the same cell. Relationship in these instances is conveniently evaluated by similarity in morphologic characteristics of the phage particles and the results of serologic tests made with them. Thus, closely related phages are characterized by particles which are very similar in size and shape and show strong cross reactions in serologic tests. In contrast to the results of mixed infection with related phages, Delbrück noted many years ago that simultaneous presentation to *E. coli* of pairs of unrelated phages resulted in blocking replication of one or the other phage. Successful infection apparently depended on which phage initiated the process first. For example, in an experiment in which the infectious mixture contained coliphages T1 and T7, about one-third of the bacteria liberated only T1 and two-thirds only T7 phages. This phenomenon was termed *mutual exclusion.* The crucial basis for such interference is still debated. There are several explanations, and one is probably more plausible than the other depending on the system.

One possibility is that the first phage to inject its DNA into a cell induces vigorous synthesis of cell wall or membrane so that entrance of DNA from the second phage is obstructed. Support for this idea is provided by the observation that if a cell is about to lyse from single infection and is then superinfected with a second related phage, the lysis is postponed for a considerable period of time. This is generally attributed to a repair and strengthening of the bacterial wall and membrane as an immediate consequence of infection, such action serving to obstruct entrance of the DNA from the second virus. Similarly, "lysis from without," a disruption of bacterial cells normally brought about by treating uninfected cells with a high multiplicity of phages, does not readily occur if it is attempted with cells in which an infectious process has already started.

If one of the pair of unrelated viruses in a mixed infection is of the T-even type, which destroys the host cell DNA soon after infection, and the other is a phage such as T1, T7, or lambda, which needs help from host DNA in order to replicate, it is clear that mixed infection will favor the T-even phage and abolish replication of the second phage. This alternative presupposes that both phages gain entrance to the same cell and hence that interference occurs intracellularly.

Another basis for intracellular interference between phages in mixed infection might involve the differential action of phage-induced deoxyribonuclease. This enzyme, presumed to be a gene product of T-even DNAs and responsible for the destruction of host DNA, might also destroy the DNA of a second unrelated phage. The DNA of T-even phages is apparently less susceptible to degradation than other DNAs, possibly because T-even DNAs are glucosylated and other DNAs are not.

Another type of interference, previously noted when lysogeny was discussed, is also called immunity. Here the mixed infection situation involves a prophage already present in the bacterial cell and the exclusion of a related, superinfecting phage. Exclusion is an intracellular phenomenon apparently based on the production of phage-specific protein called *immunity substance* or *repressor.* The repressor, by attaching to viral DNA at a crucial locus, prevents transcription of viral genome so that neither the prophage which coded for the repressor nor the superinfecting phage can replicate to form progeny virus particles.

Most of the interference phenomena observed with animal and plant viruses involve mixed infection of the type in which the components of the mixture are supplied at different times. For example, as noted earlier, cells infected with certain animal viruses induce the production and release to adjacent cells of proteins called interferons. These substances depress the multiplication of the inducing virus as well as viruses quite unrelated to the inducing virus. The precise mechanism of action of interferons is unknown, but, as discussed in the previous chapter, it may involve a specific interference with transcription, translation, or both.

Another type of interference observed with animal viruses involves the blockage or destruction of cellular receptors. For example, the presence of excess avian leukosis virus either added or derived from infected cells greatly inhibits the infection of chick embryo cells by Rous sarcoma virus. In fact, this observation provides a basis for an assay of the leukosis virus, since the chick cells show no obvious effects of the presence of leukosis virus but develop little clumps (infectious foci) in proportion to the numbers of cells infected with Rous sarcoma virus. Hence if the cells are first infected with leukosis virus and subsequently sarcoma virus is added, the number of infectious foci is reduced in proportion to the extent of infection with leukosis virus. Similarly, infection of HeLa cells with large doses of poliovirus or Coxsackie virus renders the cells resistant to attachment of homologous but not heterologous virus. This is consistent with the idea that different viruses may use different attachment sites. In all these cases, conditions known to dissociate viral coat protein from cellular receptors (such as low pH) reverse the interference and thus support the concept that interference is due to blocking of cellular receptors by residual viral material.

A striking example of interference, first investigated in detail many years ago by von Magnus, results from the inoculation of chick embryos with heavy doses of influenza virus, especially if such inoculation is repeated in series three or four times. The product of this kind of infection is virus with low infectivity but high hemagglutinating potency. It has been called "incomplete virus" and seems to be composed of a mixed population of particles, many of which are deficient in nucleic acid content. Hence the components of mixed infection are complete and incomplete virus particles, and the von Magnus effect, as it is often called, is one of autointerference. Though the mechanism of this phenomenon is not known, it is possible that it is based on a premature release of virus from the surface of infected cells, where viruses such as influenza virus mature into complete particles. Inoculation with a large dose of virus would provide excess viral neuraminidase, which has been postulated to cause a premature release of virus from the cell surface so that a greater than normal number of particles emerge without complete genomes. Occurrence of the influenza genome in several discrete segments, as is now supposed, would favor deficient assembly of viral particles in an accelerated process.

Interference between plant viruses has been used for years as a criterion of strain relationship. The phenomenon is best observed when a plant is available which responds to one virus with a systemic disease and to the second virus by development of local lesions. Such is the case with *Nicotiana sylvestris* inoculated successively with the common and yellow aucuba strains of tobacco mosaic virus. TMV causes a systemic disease in *N. sylvestris* marked by a green mosaic, whereas yellow aucuba causes only necrotic local lesions. If *N. sylvestris* is thoroughly infected with TMV and then inoculated with the aucuba strain, infection is either greatly reduced or abolished, as determined by lesion counts. On the other hand, an unrelated virus such as tobacco ringspot virus will multiply and produce its characteristic symptoms on top of the green mosaic caused by TMV.

There are some exceptions which argue against the use of this interference test as the sole evidence of strain relationship. For example, none of the strains of sugar beet curly top virus protects against other ones in water pimpernel; and conversely, interference is observed when a lesion-producing strain of TMV (*Nicotiana* virus 6) is applied to tobacco leaves previously infected with the unrelated celery mosaic, cucumber mosaic, or potato veinbanding mosaic viruses. The nature of interference in any of these mixed infections is obscure except that it is presumably intracellular rather than involving attachment or penetration.

Even more mysterious is the type of interference observed by Ross and his associates in which infection of a leaf with one virus that causes local lesions induces a resistance to the same or unrelated viruses subsequently rubbed on other leaves not containing the first virus. This systemic spread of interference is demonstrated not by a reduction in the number of lesions caused by the second virus, but rather by a substantial decrease in size of the lesions and hence in numbers of infected cells. (In general, local lesions vary in diameter from less than a millimeter to 6 to 8 mm and in number of cells

TABLE 5-9 **Interference between Viruses Causing Local Lesions When Inoculated on Different Leaves of Pinto Beans***

Viruses used		*Relative lesion size*†
First inoculation	*Second inoculation*	
Tobacco mosaic	Tobacco mosaic	55
Southern bean mosaic	Tobacco mosaic	59
Tobacco mosaic	Southern bean mosaic	44
Southern bean mosaic	Southern bean mosaic	43
Alfalfa mosaic	Tobacco mosaic	50
Tobacco mosaic	Alfalfa mosaic	50
Alfalfa mosaic	Alfalfa mosaic	55
Tobacco necrosis	Tobacco mosaic	52

* The first virus was inoculated on one primary leaf and the second on the opposite primary leaf 7 days later.

† Diameter of lesions on leaves opposite previously inoculated ones as a percentage of the diameter of lesions on leaves opposite previously uninoculated leaves.

SOURCE: Adapted from Ross, 1966.

encompassed by about 200 to 2,000). Some examples of this long-distance interference are given in Table 5-9. Such restriction of infection is suggestive of antibody or interferon action observed in animal systems. However, neither of these effects is known in plants.

References

BOOKS

Fenner, F.: *The Biology of Animal Viruses*, vol. 1, Academic, New York, 1968.

Hayes, W.: *The Genetics of Bacteria and Their Viruses*, 2d ed., Wiley, New York, 1968.

Stahl, F. W.: *The Mechanics of Inheritance*, 2d ed., Prentice-Hall, Englewood Cliffs, N.J., 1969.

Stent, G. S.: *Molecular Genetics*, Freeman, San Francisco, 1971.

JOURNAL ARTICLES AND REVIEW PAPERS

Complementation

Atabekov, J. G., N. D. Schaskolskaya, T. I. Atabekova, and G. A. Sacharovskaya: Reproduction of Temperature-Sensitive Strains of TMV under

Restrictive Conditions in the Presence of Temperature-Resistant Helper Strain, *Virology* **41**:397–407, 1970.
Bancroft, J. B.: A Virus Made from Parts of the Genomes of Brome Mosaic and Cowpea Chlorotic Mottle Viruses, *J. Gen. Virol.* **14**:223–228, 1972.
Benzer, S.: The Elementary Units of Heredity, in *The Chemical Basis of Heredity*, W. D. McElroy and B. Glass (eds.), pp. 70–93, Johns Hopkins, Baltimore, 1957.
Berry, G. P., and H. M. Dedrick: A Method for Changing the Virus of Rabbit Fibroma (Shope) into That of Infectious Myxomatosis (Sanarelli), *J. Bacteriol.* **31**:50–51, 1936.
Bol, J. F., L. Van Vloten-Doting, and E. M. J. Jaspars: A Functional Equivalence of Top Component aRNA and Coat Protein in the Initiation of Infection by Alfalfa Mosaic Virus, *Virology* **46**:73–85, 1971.
Hanafusa, H., T. Hanafusa, and H. Rubin: The Defectiveness of Rous Sarcoma Virus, *Proc. Natl. Acad. Sci. U.S.A.* **49**:572–580, 1963.
Kado, C. I., and C. A. Knight: The Coat Protein Gene of Tobacco Mosaic Virus. I. Location of the Gene by Mixed Infection, *J. Mol. Biol.* **36**:15–23, 1968.
Kassanis, B.: Satellitism and Related Phenomena in Plant and Animal Viruses, *Adv. Virus Res.* **13**:147–180, 1968.
Lister, R. M.: Tobacco Rattle, NETU, Viruses in Relation to Functional Heterogeneity in Plant Viruses, *Fed. Proc.* **28**:1875–1889, 1969.
Padgett, B. L., and J. K. N. Tomkins: Conditional Lethal Mutants of Rabbitpox Virus. III. Temperature-sensitive (*ts*) Mutants; Physiological Properties, Complementation and Recombination, *Virology* **36**:161–167, 1968.
Rapp, F., and J. L. Melnick: Papovavirus SV40, Adenovirus and Their Hybrids: Transformation, Complementation, and Transcapsidation, *Prog. Med. Virol.* **8**:349–399, 1966.
Rubin, H.: Virus without Symptoms and Symptoms without Virus; Complementary Aspects of Latency as Exemplified by the Avian Tumor Viruses, *Perspect. Virol.* **4**:164–174, 1965.
Sänger, H. L.: Defective Plant Viruses, in *Molecular Genetics*, 4 Wiss. Kong. Ges. Deutsche Ärze, Berlin, 1967, H. G. Wittmann and H. Schuster (eds.), pp. 300–336, Springer-Verlag, Berlin, 1968.
———: Functions of the Two Particles of Tobacco Rattle Virus, *J. Virol.* **3**:304–312, 1969.
Van Kammen, A.: Plant Viruses with a Divided Genome, *Ann. Rev. Phytopathol.* **10**:227–252, 1972.
Vogt, P. K.: RNA Tumor Viruses: The Problem of Viral Defectiveness, in *Viruses Affecting Man and Animals*, M. Sanders and M. Schaeffer (eds.), pp. 175–190, W. H. Green, St. Louis, 1971.

Interference

Bennett, C. W.: Interactions between Viruses and Virus Strains, *Adv. Virus Res.* **1**:39–67, 1953.

Crowell, R. L.: Specific Cell Surface Alteration by Enteroviruses as Reflected by Viral-attachment Interference, *J. Bacteriol.* **91**:198–204, 1966.

Delbrück, M.: Interference between Bacterial Viruses. III. The Mutual Exclusion Effect and the Depressor Effect, *J. Bacteriol.* **50**:151–170, 1945.

Ross, A. F.: Systemic Effects of Local Lesion Formation, in *Viruses of Plants,* A. B. R. Beemster and J. Dijkstra (eds.), pp. 127–150, Wiley, New York, 1966.

Steck, F. T., and H. Rubin: The Mechanism of Interference between an Avian Leukosis Virus and Rous Sarcoma Virus. I. Establishment of Interference. II. Early Steps of Infection by RSV of Cells under Conditions of Interference, *Virology* **29**:628–653, 1966.

Von Magnus, P.: Incomplete Forms of Influenza Virus, *Adv. Virus Res.* **2**:59–79, 1954.

Marker Rescue (Cross Reactivation)

Doermann, A. H.: The Analysis of Ultraviolet Lesions in Bacteriophage T4 by Cross Reactivation, *J. Cell Comp. Physiol.* **58**:suppl. 1, 79–93, 1961.

Kilbourne, E. D., F. S. Lief, J. L. Schulman, R. I. Jahiel, and W. G. Laver: Antigenic Hybrids of Influenza Viruses and Their Implications, *Perspect. Virol.* **5**:87–106, 1967.

Mating and Genetic Recombination

Best, R. J.: Tomato Spotted Wilt Virus, *Adv. Virus Res.* **13**:65–146, 1968.

Clark, A. J.: Toward a Metabolic Interpretation of Genetic Recombination of *E. coli* and Its Phages, *Ann. Rev. Microbiol.* **25**:437–464, 1971.

Cooper, P. D.: A Genetic Map of Poliovirus Temperature-sensitive Mutants, *Virology* **35**:584–596, 1968.

Davern, C. I.: Molecular Aspects of Genetic Recombination, *Prog. Nucleic Acid Res. Mol. Biol.* **11**:229–258, 1971.

Edgar, R. S., and R. H. Epstein: The Genetics of a Bacterial Virus, *Sci. Am.* **212**:70–78, 1965.

Hershey, A. D., and R. Rotman: Genetic Recombination between Host-range and Plaque-type Mutants of Bacteriophage in Single Bacterial Cells, *Genetics* **34**:44–71, 1949.

Hirst, G. K.: Genetic Recombination with Newcastle Disease Virus, Poliovirus, and Influenza, *Cold Spring Harbor Symp. Quant. Biol.* **27**:303–398, 1962.

Streisinger, G., R. S. Edgar, and G. H. Denhardt: Chromosome Structure in Phage T4. I. Circularity of the Linkage Map, *Proc. Natl. Acad. Sci. U.S.A.* **51**:775–779, 1964.

Multiplicity Reactivation

See Books: Fenner, chap. 9.

Luria, S. E., and R. Dulbecco: Genetic Recombinations Leading to Production of Active Bacteriophage from Ultraviolet Inactivated Bacteriophage Particles, *Genetics* **34**:93–125, 1949.

Schäfer, W., and R. Rott: Herstellung von Virusvaccinen mit Hydroxylamin. Verlauf der Inaktivierung und Wirkung des Hydroxylamins auf verschiedene biologische Eigenschaften einiger Viren, *Z. Hyg. Infektionskr.* **148**:256–268, 1962.

Phenotypic Mixing

Holland, J. J., and C. E. Cords: Maturation of Poliovirus RNA with Capsid Protein Coded by Heterologous Enteroviruses, *Proc. Natl. Acad. Sci. U.S.A.* **51**:1082–1085, 1964.

Kassanis, B., and C. Bastow: In vivo Phenotypic Mixing between Two Strains of Tobacco Mosaic Virus, *J. Gen. Virol.* **10**:95–98, 1971.

——— and ———: Phenotypic Mixing between Strains of Tobacco Mosaic Virus, *J. Gen. Virol.* **11**:171–176, 1971.

Novick, A., and L. Szilard: Virus Strains of Identical Phenotype but Different Genotype, *Science* **113**:34–35, 1951.

Rochow, W. F.: Barley Yellow Dwarf Virus, Phenotypic Mixing: and Vector Specificity, *Science* **167**:875–878, 1970.

Roizman, B., and L. Aurelian: Abortive Infection of Canine Cells by Herpes Simplex Virus. I. Characterization of Viral Progeny from Cooperative Infection with Mutants Differing in Capacity to Multiply in Canine Cells, *J. Mol. Biol.* **11**:528–538, 1965.

Streisinger, G.: Phenotypic Mixing of Host Range and Serological Specificities in Bacteriophages T2 and T4, *Virology* **2**:388–398, 1956.

CHAPTER 6
MUTATION

It appears that the genetic material (nucleic acid) of viruses is replicated with remarkable fidelity, as adduced from the constancy of observed properties of many viruses. Nevertheless, in something like one in 10^2 to one in 10^9 duplications, depending on the virus and the host, it appears that a heritable change in the nucleic acid may occur. This is called a *mutation.* It should be emphasized that the consequences of some mutations are not readily observed and that other mutations are viable only under a special set of conditions. For example, a mutant may multiply in one host (permissive host) but not in another (nonpermissive or restrictive host), whereas the parental virus multiplies in both. Likewise a "temperature-sensitive" (*ts*) mutant may replicate at a certain temperature and not at a higher temperature, while the wild type multiplies at both temperatures. For all these reasons many mutations doubtless escape notice. In addition, the spontaneously arising mutant may be overwhelmed by unmutated progeny of the parental virus and is lost unless some selective event intervenes. In short, readily detectable, spontaneous mutations of viruses occur with low frequency (a fraction of 1 percent);

and for the reasons given above, it is not possible to ascribe a quantitatively accurate figure for the general mutation rate of any virus. However, if one scores only a particular type of mutant, accurate quantitative figures may be obtainable and the rate of mutation can be analyzed mathematically (see Chap. 5 in Drake, for example).

A natural selection of mutants may occasionally occur. This happens more frequently with some viruses than with others. Such a selection becomes dramatically evident if the selected mutant happens to be more virulent and invasive than the currently common virus. An event of this kind has been blamed for the disastrous worldwide epidemic (called a *pandemic*) of influenza in 1918–1919, which is estimated to have taken more than 15 million lives and then to have vanished. New strains of influenza virus continue to appear, cause disease around the world for a few years, fade out, and are replaced by others. None has been so lethal as the 1918–1919 strain.

Various processes are used in the laboratory for selecting mutants from a population of viruses. Sometimes passage through a host that is different from the one in which the virus has been multiplying will favor the mutant. More commonly, use of techniques which give isolated colonies of virus are employed. Such pure cultures have been obtained from plaques formed by phages on lawns of bacteria, plaques of tissue culture cells infected with animal viruses, or from local lesions or other distinctive spots on leaves of virus-infected plants. Medically, the deliberate isolation of mild strains has provided material for vaccines against such diseases as smallpox, yellow fever, and poliomyelitis.

6-1 *Molecular Mechanisms of Mutation*

MUTATIONS ARE CAUSED BY NUCLEOTIDE SUBSTITUTIONS, ADDITION OR DELETION OF ONE OR MORE NUCLEOTIDES, AND REARRANGEMENT OF LARGE POLYNUCLEOTIDE SEGMENTS

It appears that a variety of viable and reproducible alterations of viral nucleic acids can occur spontaneously or be induced in the laboratory. This means that there is a significant degree of flexibility in the hereditary mechanism and that many changes readily permit survival, others are conditionally lethal (permit functioning under special conditions), and some are almost certainly lethal. The terms used to describe various types of mutational alterations have become somewhat confused among different authors working with different systems. For example, some people understand a "point mutation" to describe only substitution of one nucleotide for another, whereas others broaden the definition to include addition or deletion of one or a few nucleotides. The latter seems to be stretching a point, and it appears simpler and clearer to classify types of mutations by less ambiguous but commonly used terms. This is done in Table 6-1, in which the types of mutation are defined in terms of changes in nucleic acid and subsequent changes in a protein or proteins (coded for by the nucleic acid).

It should be recognized that though the relations between mutations and proteins summarized in Table 6-1 are important, they are not the only consequences possible. Mutations may also affect regulatory functions such as might be expressed through transcription of

TABLE 6-1 **Molecular Mechanisms of Mutation**

Type	*Change in nucleic acid*	*Change in protein coded for by nucleic acid*
Substitution	One nucleotide* is replaced by a different one	No change in protein occurs, or, more commonly, one amino acid is replaced by another (see Table 6-2)
Addition-deletion	One nucleotide (or nucleotide pair)† or a sequence of a few nucleotides is added or deleted	Several amino acids are replaced by different ones (see Fig. 6-6)
Rearrangements	Large polynucleotide segments representing one or more genes are linked in an order different from the original arrangement	Some proteins are deleted by mid-gene disruptions, and different combinations of proteins appear when rearrangement involves segments of nucleic acid from different strains of virus

* Since change in a nucleotide usually means changes in the purine or pyrimidine base, rather than in the ribose or phosphate constituents, it is common to speak of the substitution of a nucleotide as a base substitution.

† In double-stranded nucleic acid, change in a nucleotide of one strand becomes a change in a nucleotide pair as soon as the affected strand is replicated.

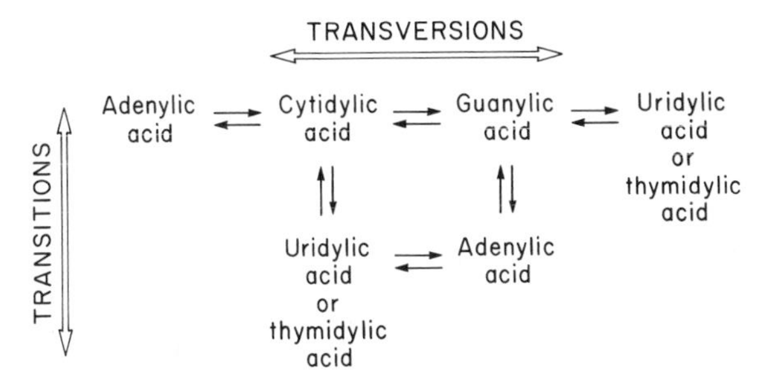

FIG. 6-1. **The possible transitions and transversions of purine and pyrimidine bases in nucleic acids.**

genes to produce special transfer RNA (tRNA) molecules. Precise relationships between such genes and their transcripts are not yet clear but could well be imagined to parallel those occurring between genes and messenger RNAs (mRNAs) and subsequently translated into proteins.

There are two common purine and two common pyrimidine bases in each of the two types of nucleic acid (DNA and RNA) that store genetic information. Theoretically, any one of the four could be replaced by any of the other three. Since the four bases occur hundreds of times in smaller nucleic acids and thousands of times in larger ones, it is clear that there are many opportunities for base substitution, and it is remarkable that substitution occurs so infrequently. In terms fairly widely used, substitution of one purine for another or one pyrimidine for another is called a *transition,* whereas substitution of a purine for a pyrimidine or the reverse is called a *transversion.* As shown in Fig. 6-1, there are four possible transitions and eight transversions.

NUCLEOTIDE SUBSTITUTIONS CAN BE EFFECTED BY RARE PROTON SHIFTS WHICH RESULT IN SUBSEQUENT MISPAIRING OF BASES

How does a nucleotide substitution occur in the intracellular replication of nucleic acids? Viral and other nucleic acids appear to be duplicated by a process in which a strand of parental nucleic acid serves as a template for synthesis of a complementary strand, as outlined in Chap. 4. It is presumed that a proper order of nucleotides is maintained by specific hydrogen bonding between the purine and pyrimidine bases of the nucleotides in the template strand and those of successive incoming nucleotides forming the nascent complementary strand. For example, in DNA synthesis an incoming thymidylic acid molecule normally pairs through its thymine with the adenine of the template strand, as indicated in Fig. 6-2*a*. However, if a hydrogen atom shifts from N-1 to the oxygen on C-6, an event that happens only rarely and transiently under physiologic conditions, then thymine could pair with guanine, as shown in Fig. 6-2*b*. Once such a substitution has occurred through this kind of mispairing, it will be perpetuated unless or until some other rare event intervenes.

Tautomeric shifts of protons like that illustrated in Fig. 6-2 for thymine or amino-imino shifts of protons can also occur occasionally with the other pyrimidine and purine bases found in nucleic acids. Such structures provide the major theoretic basis for mispairing leading to nucleotide substitution and mutation.

NUCLEOTIDE SUBSTITUTIONS CAN BE CAUSED BY MUTAGENIC CHEMICALS WHICH CONVERT ONE BASE INTO ANOTHER, CAUSE STRUCTURAL CHANGES RESULTING IN MISPAIRING OF BASES, OR OCCASIONALLY INDUCE DELETIONS

Treatment of viruses or their isolated nucleic acids with certain chemicals, or introducing specific reagents into infected cells, may also cause mutation by the substitution mechanism. The list of such mutagenic chemicals includes nitrous acid, purine and pyrimidine analogues, hydroxylamine, and various alkylating agents (alkyl sulfates, mustard gas derivatives, N-nitroso compounds, etc.).

Nitrous acid effects a direct substitution of nucleotides in nucleic acid chains by oxidative deamination of the purines and one of the pyrimidine bases. The conversion of cytosine to uracil by this means is illustrated in Fig. 6-3. The cytosine-to-uracil transition is perpetuated if the substrate for nitrous acid was single- or double-stranded RNA, but if the substrate was either single- or double-stranded DNA, two cycles of replication will result in the replacement

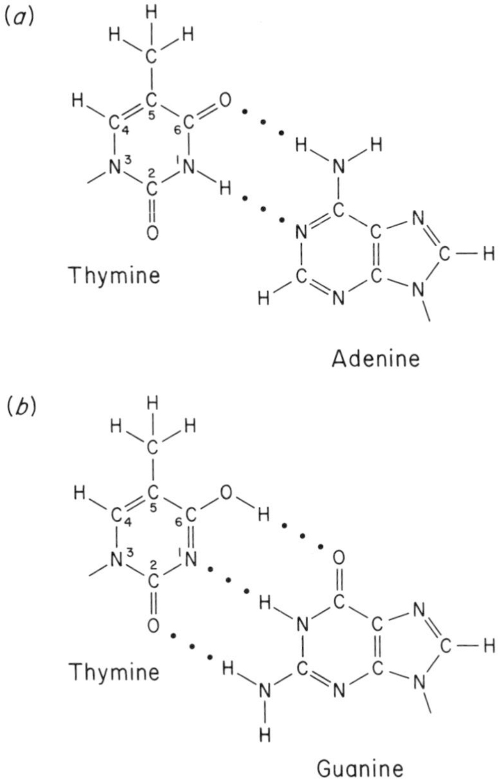

(*a*)

Thymine

Adenine

(*b*)

Thymine

Guanine

FIG. 6-2. **Normal and abnormal base pairing of thymine. (a) Hydrogen bonding between adenine and thymine with the latter in the keto form, which is the predominate structure in aqueous solution at physiologic pH values. (b) Hydrogen bonding between guanine and thymine when the latter is in the less common enol form.**

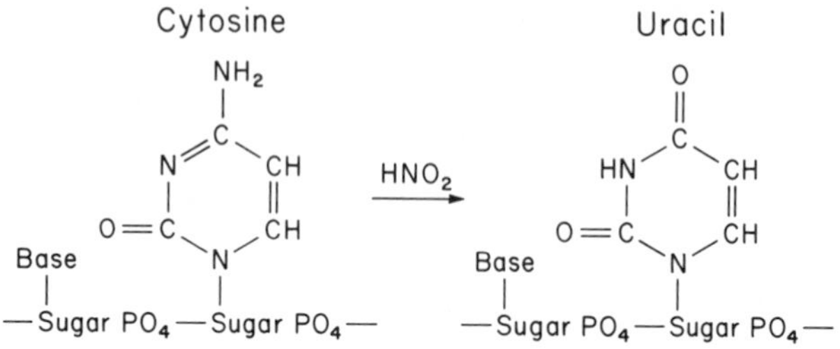

Cytosine $\xrightarrow{HNO_2}$ Uracil

Cytosine: NH_2 / C / N CH / O=C CH / N / Base / —Sugar PO_4—Sugar PO_4—

Uracil: O / C / HN CH / O=C CH / N / Base / —Sugar PO_4—Sugar PO_4—

FIG. 6-3. **Oxidative deamination of cytosine in a polynucleotide chain to yield uracil.**

of cytosine with thymine. Similar deaminations of adenine and guanine yield hypoxanthine and xanthine, respectively. The latter are naturally occurring purines but, unlike uracil, are not commonly found in nucleic acids. However, hypoxanthine can pair with cytosine, which in turn pairs with guanine. Thus, two cycles of replication after action of nitrous acid on adenine result in the substitution of guanine for adenine. Xanthine, produced by the deamination of guanine, can pair with cytosine, which in the next replicative step will pair with guanine, giving back the original combination. However, xanthine is said to ionize more at pH values around neutrality than other bases, and this interferes with its base-pairing function and hence tends toward a lethal effect.

In short, the major mutagenic effect of nitrous acid is associated with its action on adenine and cytosine to give the transitions A to G and C to T (U), or in terms of base pairs, AT to GC and GC to AT. Since these two transitions are the reverse of one another, it might be supposed that nitrous acid–induced mutations could be reverted by nitrous acid treatment (or by 5-bromouracil, an in vivo transition-inducing agent). Such reversions have in fact been reported. Less well-defined effects attributed to nitrous acid are deletion mutations (observed in phage T4) and cross-linking of DNA strands; the latter is perhaps related mechanistically to the former.

Nitrous acid deaminates bases in both free nucleic acid and nucleic acid ensheathed in the viral protein coat. However, the rate of reaction may be an order of magnitude slower in the reaction with intact virus, presumably because of side reactions of the nitrous acid with viral protein and difficulties in penetrating the protein coat. In addition, the reaction of the bases in whole virus particles may be modified, as it was shown to be with tobacco mosaic virus. In isolated TMV RNA, adenine is deaminated 50 percent faster than cytosine, whereas in whole virus it is just the reverse.

Hydroxylamine and methylating agents cause alterations in bases leading generally to transitions when the nucleic acids are replicated. For example, hydroxylamine is thought to react rather specifically at pH 6 with cytosine to produce one or all of the N-hydroxylated derivatives shown in Fig. 6-4. The first and third of the compounds shown in Fig. 6-4 or their tautomers form hydrogen bonds with adenine, rather than with guanine, as cytosine normally does. The end result, upon replication of the nucleic acid, is the substitution of thymine (or uracil) for cytosine. At pH 9, the reaction of hydroxylamine with uracil, which tends to be lethal, is about eight times as fast as the reaction with cytosine. At either pH 6 or 8, hydroxylamine, in contrast to nitrous acid, seems to act only on isolated RNA.

A type of mispairing culminating in nucleotide substitution can occur if base analogues such as 5-bromo-, 5-chloro-, or 5-fluorouracil

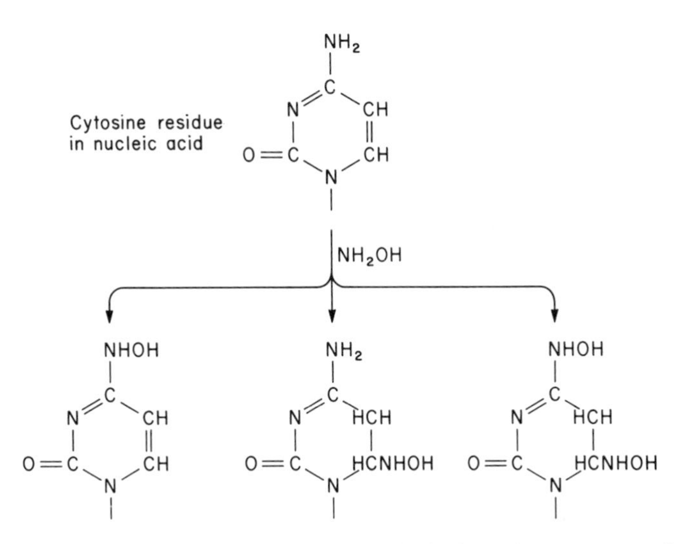

FIG. 6-4. **Reaction of cytosine with hydroxylamine to produce derivatives with mutagenic potentiality.**

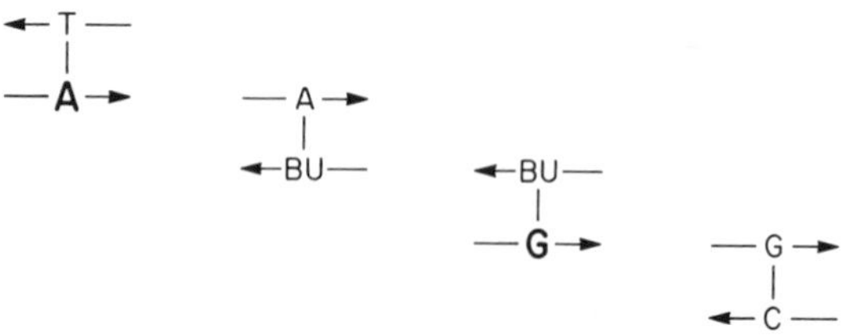

FIG. 6-5. **A to G transition in DNA (AT to GC in terms of base pairs), brought about by introduction of 5-bromouracil (BU). The arrows indicate orientation of polynucleotide chains. The composition of base pairs at a given locus is followed from left to right through three successive replications of the DNA, starting with insertion of BU in place of T in the first step.**

are present in the cell when nucleic acids are replicating. For example, owing to tautomerism, which probably occurs with a higher frequency with bromouracil (BU) than with thymine, BU can pair with either adenine or guanine and hence can, by its incorporation in the nucleotide chain, lead to substitution of guanine for adenine, as shown in Fig. 6-5.

In whatever manner it is brought about, a nucleotide substitution in a viral or other nucleic acid may or may not result in an altered protein. This can be illustrated by the mispairing examples given in Table 6-2, in which adenine is shown paired with cytosine rather than with thymine. In the first case (left side of Table 6-2), the nucleotide substitution, while changing the codon, still leads to the same amino acid in the protein, namely, glycine (see Table 6-3 for the codon catalog). In the second case (right side of Table 6-2), the nucleotide substitution changes not only the codon but also the amino acid. A general point is illustrated by this example, namely, that substitution of the first or second nucleotide of a codon triplet (reading from the left) often leads to a different amino acid in the resultant gene product (protein), whereas substitution of the third nucleotide frequently yields the same amino acid (see Table 6-3). It should also be noted that mispairing could occur at points other than the ones indicated in Table 6-2 in the sequence of parental DNA → replicated DNA → mRNA. Also, if the viral nucleic acid were either single-stranded or double-stranded RNA, rather than DNA, there would be opportunities for the same kind of mispairing.

However, even when nucleotide substitution leads to a different

TABLE 6-2 **Two Examples of Normal Base Pairing and Mispairing of Adenine in the Sequence Leading from Parental DNA Strand to Protein**

Sequence	*Normal pairing*	*Mispairing*	*Normal pairing*	*Mispairing*
Parental DNA plus* strand	– CCA –	– CCA –	– TAG –	– TAG –
	. . .	. . .	. . .	. . .
Complementary DNA minus* strand	. . . – GGT –	. . . – GG[C] –	. . . – ATC –	. . . – A[C]C –
New DNA plus strand	– CCA – . . .	– CCG – . . .	– TAG – . . .	– TGG – . . .
Messenger RNA	. . . – GGU –	. . . – GGC –	. . . – AUC –	. . . – ACC –
Amino acid specified	Glycine	Glycine	Isoleucine	Threonine

* One strand of DNA is arbitrarily designated the "plus" strand, and its complementary strand, either in the original duplex or arising on the plus strand template during replication, is called the "minus" strand. The points in replication at which mispairing takes place are indicated by boxes. It will be recognized that mispairing could occur at other points, e.g., in transcription of mRNA from the new DNA plus strand.

amino acid in the protein, the properties of the protein may not be significantly altered. Direct indication of such an alteration depends on whether the particular amino acid involved is crucial in the conformation and hence biologic activity of the protein in question.

In considering the codon catalog (Table 6-3) and the potentialities for mutational change by nucleotide substitution, it should be noted that every conceivable codon has meaning, despite the unfortunate term "nonsense" which has been employed in some quarters. All nucleotide substitutions which can be translated into functional proteins fall potentially into a "missense" category. However, the codons which have been called "nonsense" codons (UAG, UAA, and UGA) are more accurately designated "chain-termination" codons, since, when they appear in mRNA, this is the function they seem to serve, at least in bacterial systems. In fun, UAG was termed "amber," UAA, "ochre," and UGA, "opal" or "umber." These terms, especially "amber," are used a fair amount to describe mutants in which an alteration of the nucleic acid has introduced a chain-termination codon where there previously was none. Synthesis of a specific protein is thereby aborted.

TABLE 6-3 **Genetic Code***

Ala	GCA	Gly	GGA	Pro	CCA
	GCC		GGC		CCC
	GCG		GGG		CCG
	GCU		GGU		CCU
Arg	AGA	His	CAC	Ser	AGC
	AGG		CAU		AGU
	CGA				UCA
	CGC	Ile	AUA		UCC
	CGG		AUC		UCG
	CGU		AUU		UCU
Asp	GAC	Leu	CUA	Thr	ACA
	GAU		CUC		ACC
			CUG		ACG
Asn	AAC		CUU		ACU
	AAU		UUA		
			UUG	Trp	UGG
Cys	UGC				
	UGU	Lys	AAA	Tyr	UAC
			AAG		UAU
Glu	GAA				
	GAG	Met	AUG	Val	GUA
					GUC
Gln	CAA	Phe	UUC		GUG
	CAG		UUU		GUU

NOTE: Termination codons: amber (UAG), ochre (UAA), opal or umber (UGA); initiation codons: (AUG), (GUG).

* For a brief explanation of the derivation of the genetic code, see H. A. Sober (ed.), *Handbook of Biochemistry*, The Chemical Rubber Publishing Co., Cleveland, I:90, 1968.

SOME SPONTANEOUS MUTATIONS OF VIRUSES HAVE BEEN ATTRIBUTED TO ADDITIONS AND DELETIONS OF NUCLEOTIDES WHICH OCCUR OCCASIONALLY DURING REPLICATION AND RECOMBINATION. ADDITIONS AND DELETIONS CAN ALSO BE CAUSED EXPERIMENTALLY BY ADDING INTERCALATING CHEMICALS DURING REPLICATION OF VIRAL NUCLEIC ACID

The view has been advanced by Brenner and associates and supported by some data from coliphage experiments that many spontaneous mutations of the T4 phage can be attributed to addition or

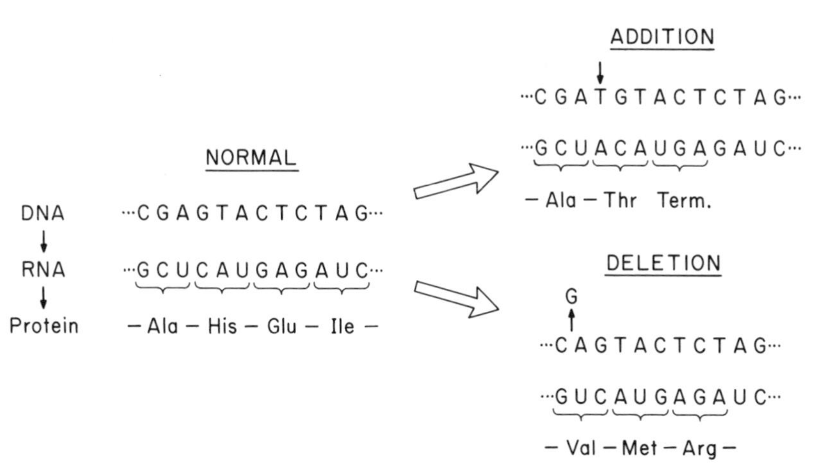

FIG. 6-6. **Examples of the effects on the coding mechanism of the addition or deletion of a single nucleotide.**

deletion of nucleotides. Since transcription of the genetic code appears to begin at a fixed point and to move along a linear series of nonoverlapping, nucleotide triplets, addition or deletion of even a single nucleotide can have a profound effect on the transcription process. The transcribing is shifted so that every codon to the right of the addition or deletion is altered; this is shown in Fig. 6-6, by examples using small segments of DNA and RNA. Mutants presumably produced from such altered transcription are called *frame-shift* or *phase-shift* mutants, because the frame of reference for transcribing the nucleotides as triplets has been shifted or the transcription has been shifted out of phase. Such mutants will be characterized by a defective protein, either because the frame shift creates a spurious termination codon, thus shortening the peptide chain, as shown in the example of nucleotide addition given in Fig. 6-6, or because the frame shift greatly changes the composition of the resultant peptide chain, as shown by a comparison of the segments of normal and deletion peptides in Fig. 6-6.

The consequences of deletion can be mitigated if the deletion is followed by an equivalent addition; i.e., if one nucleotide (or nucleotide pair) has been deleted, the proper frame of reference can be restored if one nucleotide (or nucleotide pair) is added. Such a situation was found by Terzaghi and associates in a comparison of phage lysozymes produced by wild type T4 and a double mutant induced by proflavine. The relevant data are presented in Table 6-4.

Likewise, potential gross alterations initiated by addition can be

TABLE 6-4 **Changes in Sequence of Amino Acids Observed in the Lysozyme of a Proflavine Mutant of T4 Coliphage together with a Suggested Combination of Deletion and Addition of Single Nucleotides in the Appropriate Segment of mRNA that Would Account for the Observed Differences**

	Some amino acid sequences in lysozyme							
Wild type	–Thr –	Lys –	Ser –	Pro –	Ser –	Leu –	Asn –	Ala –
Mutant	–Thr –	Lys –	Val –	His –	His –	Leu –	Met –	Ala –
	mRNA Codons							
Wild type	–ACA	AAA	AGU	CCA	UCA	CUU	AAU	GCU –
Delete third nucleotide (A) in first codon and shift succeeding nucleotides one position to the left to form new triplets. Add G just before last two nucleotides. The resulting messenger:								
Mutant	–ACA	AAA	GUC	CAU	CAC	UUA	AUG	GCU –

SOURCE: Adapted from Terzaghi et al., 1966.

avoided by an equivalent deletion. Conceptually at least, the addition or deletion of nucleotides in the form of one or more complete codons might be far less drastic in effect than addition or deletion of a single nucleotide, because frame shifts would be avoided.

Massive addition or deletions can occur by breakage and reunion processes such as might occur when a phage genome is released from a lysogenized bacterial cell. Perfect release restores the phage genome essentially intact, but splitting out the phage DNA by breaks in the wrong places results in deletion of some of the phage nucleic acid and addition of a segment of the bacterial DNA.

The frame-shift type of mutation has been induced in bacteriophage DNA, especially T4, by treating infected *E. coli* with acridine (Fig. 6-7), or, better yet, with various amino acridines (Table 6-5). As indicated in Table 6-5, the more basic acridine derivatives tend to be more mutagenic in terms of amount needed for optimal mutagenic action and, to a fair extent, in capacity to cause mutation (last column of Table 6-5). However, evaluation of the mutagenicity of the acridines is complicated by other factors, such as differences in ability of different acridines to penetrate cells and to interact with DNA. Also, acridines can induce mutations photodynamically, so that the presence or absence of visible light becomes a factor. However, the mutations

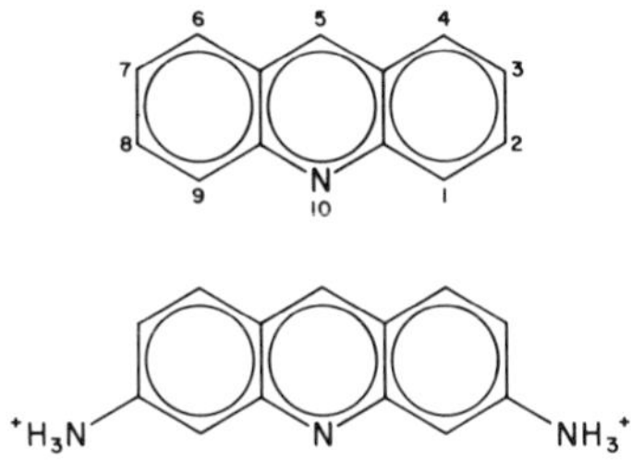

Acridine

Proflavine
(2,8-Diaminoacridine)

FIG. 6-7. **The acridine nucleus and proflavine, an acridine derivative mutagenic for some bacteriophages.**

resulting from photodynamic function of acridines appear to be of the nucleotide substitution type, rather than addition-deletion. The intracellular mutagenesis of poliovirus by proflavin is probably an example of this type of effect.

The precise mechanism of action of acridines and similar mutagenic agents is not known. However, considerable evidence, including physical chemical data and Cairns's autoradiographic visualization of chain elongation of coliphage T2 DNA by proflavin, supports the idea that acridines are intercalated (wedged in) between adjacent base pairs. Such intercalated molecules are, in the Streisinger model, responsible for stabilizing mispairing that occurs between complementary strands of DNA which have undergone breakage and partial

TABLE 6-5 **Relative Mutagenicity of Some Acridines for T4 Bacteriophage**

Compound	*pK*	*Optimal mutagenic concentration, μg/ml*	*Mutagenicity*
1-Aminoacridine	4.2	150	Weak
Acridine	5.3	90	Moderate
3-Aminoacridine	5.6	80	Weak
4-Aminoacridine	5.7	80	Strong
2-Aminoacridine	7.7	16	Strong
2,7-Diaminoacridine	7.8	16	Moderate
2,8-Diaminoacridine	9.3	5	Strong
5-Aminoacridine	9.6	8	Strong
2,5-Diaminoacridine	11.1	8	Strong

SOURCE: Adapted from Drake, 1970.

strand separation. Mispairing is accompanied by looping out of one of the strands of the duplex, with or without some excision by nuclease, and is followed by repair which leaves one strand with its original composition while the other is either lengthened or shortened by one or more nucleotides. These steps are illustrated in Fig. 6-8.

Rearrangements involving large segments of viral genome probably occur with all types of viruses during mixed infection with different strains of virus. This has been discussed in Chap. 5 under Mating as an example of genetic recombination. Since recombinants issuing from mating show stability of marker characteristics in succeeding generations, they fulfill the definition given earlier for mutants, as well as exemplifying rearrangement. There are indications, at least with phages, that intragenomic as well as intergenomic exchanges occasionally occur during viral replication. In an extreme case, for example, so many extra nucleotide pairs were inserted as to markedly increase the densities of the phages concerned.

RADIATIONS CAN CAUSE MUTATIONS AT A LOW FREQUENCY, BUT THE MECHANISM IS UNCERTAIN

The mutagenic action of radiations on various eucaryotic organisms has been a widely investigated field since Muller's pioneer investigations over 40 years ago. The mutagenesis of viruses by radiations has been established more recently. In all cases the molecular mechanism of radiation-induced mutations remains somewhat obscure.

The radiations employed in studies on viruses include x-rays, gamma rays, and ultraviolet light. The latter has probably been used most abundantly because of the ready availability of sources in the form of inexpensive germicidal lamps producing strong emission of ultraviolet light at a wavelength of 253.7 nm, where the purine and pyrimidine bases of nucleic acids absorb strongly. Mutagenesis by treatment with ultraviolet light has been observed mainly with coliphages. In some cases irradiation of the virus must be accompanied by irradiation of the host cells as well in order to produce mutants. In extensive studies made on T4 coliphage, which does not require

FIG. 6-8. **Hypothetic mechanism for the production of frame-shift mutants by strand breaking and mispairing, according to the Streisinger model. As shown, after a single break has been introduced in one of the strands of DNA, strand separation occurs over a small distance, either followed by nuclease action on the open strand (right side of figure) or not (left side of figure). In both cases, looping out of a strand occurs next but**

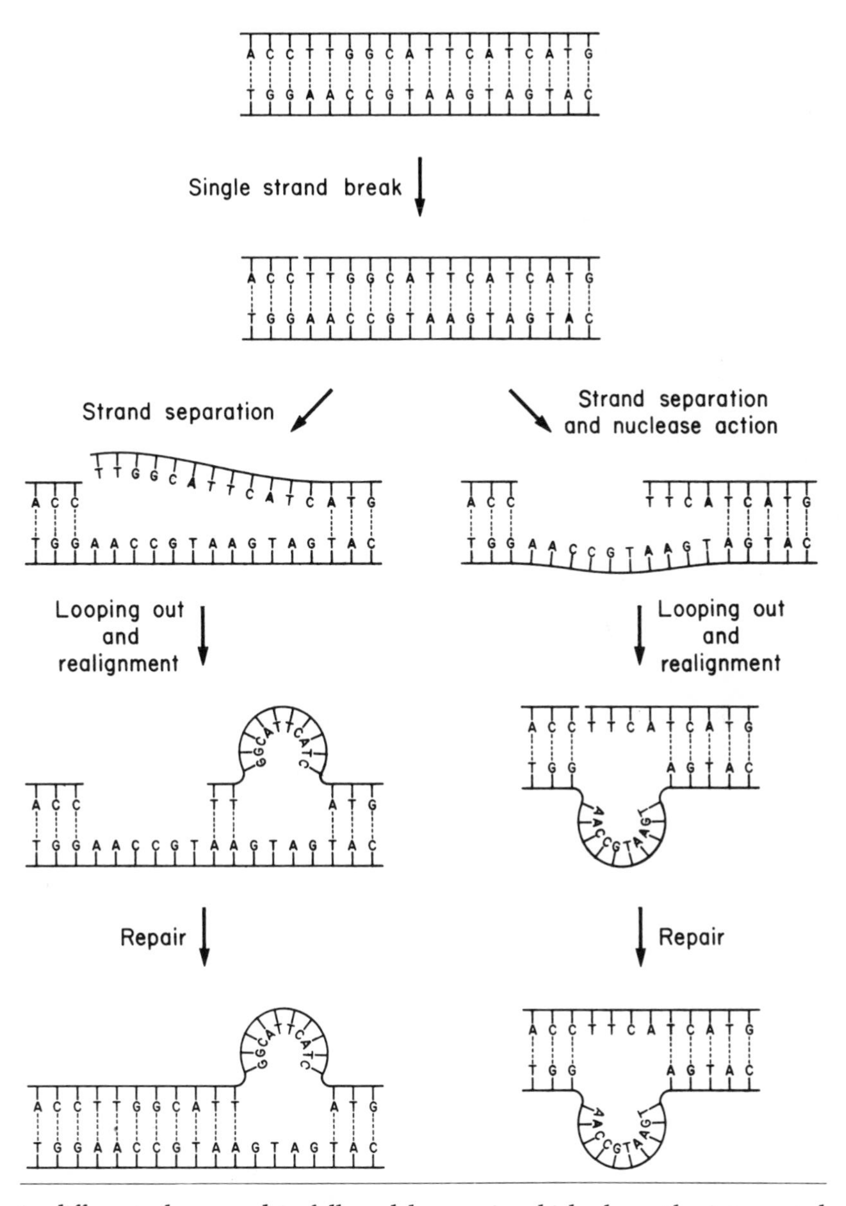

in different places, and is followed by repair which closes the interrupted strand. In the next round of replication, it is evident that the top strand of the product on the left will give rise to a duplex containing an addition, whereas the top strand of the product on the right will yield a duplex with a deletion.

irradiation of the host cell, it appears that about half the mutations induced in the rII region of the viral genome are of the frame-shift type, since they are reverted by proflavin, a potent addition-deletion reagent. Nearly all the rest of the mutations are reverted by base analogues, indicating that they probably contain nucleotide substitutions. How either of these types of mutation is brought about by irradiation with ultraviolet light is a mystery, but there are several hypotheses.

A favorite notion is based on the observation that one effect of ultraviolet irradiation, often lethal, is to promote bond formation between pyrimidine bases adjacent to each other on the same polynucleotide strand. When two thymine residues are linked, the combination is referred to as a *thymine dimer*. The lethal effect is thought to be based on the inability of RNA polymerases to transcribe regions of a gene containing such dimers. The mutation reaction might be attributed to local distortion of the DNA strand, so that copying errors would occur when the dimer-containing strand is used as a template in replication.

Cells commonly contain repair systems for maintaining the integrity of nucleic acids, especially DNA. For example, there is evidence in bacterial systems of enzymes capable of excising thymine dimers. After such excision, a DNA polymerase inserts new nucleotides, which are joined by a ligase enzyme. It has been suggested that mistakes may be made in patching the DNA strand in the manner just outlined and that those errors are the basis of the mutation.

A number of dyes, such as proflavin, methylene blue, thiopyronin, and psoralen, have mutational potentiality resulting from possible intercalation between nucleotide pairs. In addition these dyes possess the capacity to transmit sufficient energy from absorption of white light to cause inactivation or mutation. Mutations of coliphage T4 have been obtained by this means. Since such mutants can be reverted by a variety of treatments, the nature of the photodynamic alterations is assumed to include both nucleotide substitutions and frameshifts.

In general, it can be concluded that mutagenesis by whatever agent tends to be mechanistically complex and usually much better documented by genetic tests than by direct chemical evidence. This is primarily because it is much easier to get genetic data than precise chemical data. As noted earlier, the frequency of mutation, whether spontaneous or induced, is usually very low. Lethal effects generally surpass mutagenic consequences. This is demonstrable in induced mutations and assumed in the case of spontaneous mutations.

6-2 Chemical Correlation between Viral Genes and Their Protein Products

ANALYSES OF PROTEINS OF VIRAL MUTANTS SUPPORT THE THEORY OF A UNIVERSAL CODE AND ARE CONSISTENT WITH COMMON MUTATIONAL MECHANISMS

The first evidence that a chemical change in the virus can accompany mutation of that virus was the demonstration by Knight and Stanley of significantly different protein compositions for wild type and spontaneous mutants of tobacco mosaic virus. These results, together with later confirming data, are summarized in Table 6-6. It is clear that the coat proteins of these two strains of TMV differ markedly in content of aromatic amino acids. Later, Knight found many more differences in amino acid content, and finally the proteins of both strains were sequenced, with the results shown in Fig. 6-9. It should be noted that the coat protein of the HR strain differs most from TMV protein of the 300 or so strains analyzed. It is an extreme case, the opposite extreme being represented by numerous instances, of which the masked (M) strain is the prototype, where there is apparently no difference at all in the proteins. The evolutionary origin of HR is obscure. However, if it is assumed that HR evolved from TMV, it is clear from the sequential differences in their coat proteins (summarized in Fig. 6-9) that many mutational events must have occurred. Furthermore, it can be surmised on the basis of the genetic code (Table 6-3) and the nature of the amino acid exchanges observed (Fig. 6-9) that a rich variety of mutational events took place including transitions, transversions, frame shifts, and deletions. There may also have been rearrangements involving genetic recombination.

TABLE 6-6 **Aromatic Amino Acids in Two Strains of Tobacco Mosaic Virus**

	*1941 results, % amino acid**			*1962–1963 results, residues per protein subunit*†		
Strain	*Tyr*	*Trp*	*Phe*	*Tyr*	*Trp*	*Phe*
Tobacco mosaic (wild type)	3.8	4.5	6.0	4	3	8
Holmes's ribgrass (HR)	6.4	3.5	4.3	7	2	6

* See Knight and Stanley, 1941.

† See Knight, 1963.

5 10 15
Ac - Ser - Tyr - **Asn** - Ile - Thr - **Asn** - **Ser** - **Asn** - Gln - **Tyr** - **Gln** - **Tyr** - **Phe** - **Ala** - **Ala** - **Val** - Trp -
20 25 30 35
Ala - **Glu** - Pro - **Thr** - **Pro** - **Met** - **Leu** - Asn - **Gln** - CyS - **Val** - **Ser** - Ala - Leu - **Ser** - **Gln** - **Ser** - **Tyr** -
40 45 50
Gln - Thr - Gln - **Ala** - **Gly** - Arg - **Asp** - **Thr** - Val - **Arg** - **Gln** - Gln - Phe - **Ala** - **Asn** - **Leu** - **Leu** - **Ser** -
55 60 65 70
Thr - **Ile** - **Val** - **Ala** - **Pro** - **Asn** - **Gln** - Arg - Phe - Pro - Asp - **Thr** - **Gly** - Phe - **Arg** - Val - Tyr - **Val** -
75 80 85
Asn - **Ser** - Ala - Val - **Ile** - **Lys** - Pro - Leu - **Tyr** - **Glu** - Ala - Leu - **Met** - **Lys** - **Ser** - Phe - Asp - Thr -
90 95 100 105
Arg - Asn - Arg - Ile - Ile - Glu - **Thr** – Gln - **Glu** - **Glu** - **Ser** - **Arg** - Pro - **Ser** - **Ala** - **Ser** - Gln - **Val** -
110 115 120 125
Ala - Asp - Ala - Thr - **Gln** - Arg - Val - Asp - Asp - Ala - Thr - Val - Ala - Ile - Arg - Ser - **Gln** - Ile -
130 135 140
Glu - **Leu** - Leu - **Leu** - **Asn** - Glu - Leu - **Ser** - **Asn** - **His** - **Gly** - Gly - **Tyr** - **Met** - **Asp** - Arg - **Ala** -
145 150 155
Gln - Phe - Glu - - - - - **Ala** - **Ile** - - - - - Leu - **Pro** - Trp - Thr - **Thr** - **Ala** - Pro - Ala - Thr

FIG. 6-9. **Amino acid sequence of the coat protein of the Holmes ribgrass (HR) strain of tobacco mosaic virus showing in bold type the differences between HR and common TMV. Note deletions at positions 146 and 149. Compare with Fig. 3-9. (Adapted from the data of Funatsu and Funatsu, 1968, and Hennig and Wittmann, 1972.)**

Mechanistically, most of these postulated codon changes are of the type which are probably effected by anomalies in the replication of the viral nucleic acid, rather than by modification of bases in completed nucleic acid. For example, there are only about five instances in which the observed change is explicable by a cytosine-to-uracil transition such as might be effected by oxidative deamination of cytosine *in situ*.

The HR strain is also exceptional in being one of very few of the more than 300 strains of TMV analyzed that has fewer than 158 amino acids in its coat protein. It appears, therefore, that somewhere in the evolution of HR a deletion of two codons took place, so that HR has 156 amino acids rather than 158, which seems to characterize many of the strains of TMV. The deletion occurs near the C terminus of the polypeptide chain, which probably accounts for its failure to yield a denatured protein. In this connection, Wittmann-Liebold and Wittmann found upon sequencing four strains of TMV whose protein compositions were different in many respects that there were two rather large regions of homology in amino acid sequence shared by all four strains. These invariant regions, one consisting of 8 amino acids in positions 87 to 94 and another of

10 amino acids in positions 113 to 122, probably represent regions necessary for maintaining the chain folding essential to the protein in order for it to serve as a subunit in the coat of the virus.

Despite the great many differences between the coat protein of common TMV and the HR strain, it should be noted that there are also striking similarities. For example, both proteins begin the same, with acetyl-Ser-Tyr-, and both end the same, with Pro-Ala-Thr. Also, both share the feature common to many strains of TMV of having no basic amino acid in the first 40 residues at the amino terminal end of the polypeptide chain.

The production of mutants of TMV in the laboratory has been illuminating with respect not only to the structure of these viruses but also to characterization of the genetic code. Mutants have been obtained with many of the reagents listed earlier in this chapter, but nitrous acid has been by far the most productive mutagenic agent. A compilation has been made (Table 6-7) of amino acid exchanges found in a series of nitrous acid–induced mutants of TMV. These data are representative of much more extensive analyses except for omission of the large percentage of mutants not showing changes in the viral coat protein. Clearly many different amino acids are involved, and the nature of the amino acid exchanges in relation to the genetic code is consistent with the operation of that code in the plant system. This is evident from the fact that in most cases the amino acid exchange can be explained by the change in a single nucleotide, that change being of the type expected in nitrous acid mutagenesis, namely, A $\rightarrow$ G or C $\rightarrow$ U. The few exceptions noted might well represent anomalies or spontaneous mutants. Although not brought out in the table, most of the nitrous acid mutants showing any amino acid exchange showed only one. This can be taken as evidence that the genetic code is nonoverlapping, since overlapping would often bring about two exchanges simultaneously which would be next to each other in the polypeptide chain. As already noted most of the exchanges were single ones, and when more than one exchange was found they were never contiguous.

MUTATIONS CAN AFFECT THE STRUCTURE AND FUNCTION OF VIRAL PROTEINS

The nitrous acid mutants of TMV have also been useful in illustrating the consequences of mutation on the structural functioning of protein in viral protein coat and its effect on serologic specificity of

TABLE 6-7 **Amino Acid Exchanges Found, and Corresponding Codon Changes Associated with Coat-Protein Mutants Obtained by Treatment of Tobacco Mosaic Virus with Nitrous Acid**

No. of mutants	*Amino acid exchange*	*Corresponding codon change*	*Nature of change*
1	Arg → Cys	CGU → UGU	C → U
4	Arg → Gly	AGG → GGG	A → G
1	Arg → Lys	AGA → AAA	A → G
4	Asn → Ser	AAU → AGU	A → G
1	Asp → Gly	GAU → GGU	A → G
1	Gln → Arg	CAG → CGG	A → G
1	Glu → Asp	GAA → GAU	A → U
2	Glu → Gly	GAA → GGA	A → G
1	Ile → Met	AUA → AUG	A → G
5	Ile → Val	AUU → GUU	A → G
4	Pro → Leu	CCU → CUU	C → U
2	Pro → Ser	CCC → UCC	C → U
1	Ser → His	UCU → CAU	UC → CA
3	Ser → Leu	UCG → UUG	C → U
5	Ser → Phe	UCU → UUU	C → U
2	Thr → Ala	ACU → GCU	A → G
6	Thr → Ile	ACU → AUU	C → U
2	Thr → Met	ACG → AUG	C → U
1	Tyr → Cys	UAC → UGC	A → G
1	Val → Met	GUG → AUG	G → A

SOURCE: Compiled from Hennig and Wittmann, 1972.

the viral protein. Some of the nitrous acid mutants are difficult to pass in series in plants, and this was found to be related at least in part to the defective nature of the viral coat protein. Siegel and Zaitlin and their associates found different degrees of defectiveness in such mutants. In one case, the protein was so altered that it could not ensheath the viral RNA, which consequently was readily destroyed by nuclease when attempts were made to isolate or transfer the strain. Another mutant, called PM2, produced a protein which formed bizarre, open helical structures by itself but could not form a coat around the viral RNA. Analysis of the protein of PM2 showed only two exchanges, a replacement of threonine in position 28 by isoleucine (see Fig. 3-9) and an exchange of glutamic acid in position

95 for aspartic acid. It is not known whether both or only one of these two replacements is responsible for the abnormal aggregation of the PM2 coat-protein subunits. It has been suggested that the change from threonine to isoleucine is the crucial one, since this results in replacing an amino acid residue possessing a hydrophilic side chain with one of hydrophobic character. In the other exchange, the amino acids involved are very similar.

Two nitrous acid mutants of TMV studied by Sengbusch and Wittmann and called Ni118 and Ni1927 were found to have undergone the same amino acid exchange in their coat protein, namely, proline to leucine. Though this was the only exchange detected, it occurred in position 20 (see Fig. 3-9) in Ni118 and in position 156 in Ni1927. When whole particles of these two mutants were compared with particles of common TMV by serologic tests, it was found that Ni1927 could be distinguished from TMV but that Ni118 could not. From earlier studies it was known that the C-terminal ends (residue 158) of the coat-protein subunits are on the exterior of the virus particles, while the amino terminal ends are folded in toward the interior of the particles. It is clear from this example that exchange of a single amino acid can change the serologic specificity of a virus particle, especially if the exchange occurs near the surface of the particle. Incidentally, the replacement of proline in position 156 with leucine changes not only the serologic specificity of the virus particles but also their susceptibility to an enzyme. Carboxypeptidase catalyzes the hydrolytic removal of C-terminal threonine and only threonine from the coat protein of common TMV. Termination of enzyme action is due mainly to the tendency of the enzyme not to split prolyl bonds. The C-terminal sequence of the TMV coat protein is -pro-ala-thr, and hence enzymatic action essentially stops after removal of threonine. In Ni1927 the C-terminal sequence is -leu-ala-thr, and carboxypeptidase removes all three of these residues before stopping (apparently enzymatic action stops there because of a steric block, since several more amino acids are removed if isolated protein rather than whole virus is used as the substrate).

Some of the features observed in analyses of mutants in the TMV series have also been noted with bacterial viruses. The complete amino acid sequence is known for the T4 phage lysozyme, which, as noted earlier in this chapter, provides a basis for comparison with the lysozymes of mutants. There are numerous other proteins in these complex phages which can be similarly analyzed and the results related to genetic analyses.

SOME PROGRESS HAS BEEN MADE IN DIRECTLY RELATING SEQUENCES OF NUCLEOTIDES IN VIRAL GENES WITH SEQUENCES OF AMINO ACIDS IN GENE PRODUCTS

A series of small, spheroidal, RNA-containing phages such as f2, fr, M12, MS2, R17, and β is also furnishing good material for relating structure and function. These viruses are serologically related, and considerable progress has been made in determining the sequence of amino acids in their coat proteins and to a somewhat lesser degree in sequencing their RNAs. As in the TMV series, some of these RNA phages appear to have identical coat proteins, others differ by a few amino acid exchanges, and some (such as f2 and fr) display numerous differences in amino acid sequences along with marked regions of homology (Fig. 6-10). A still more distant relationship appears to exist between f2 and Qβ phages. Qβ coat protein has 131 amino acid residues per subunit, rather than the 129 in f2, and is lacking three amino acids which are present in f2, namely, histidine, methionine, and tryptophan. Nevertheless, if the amino acid sequences of these two proteins are aligned in a way that allows insertions and deletions, 30 residues occupy identical positions and

1 5 10 15
fr Ala - Ser - Asn - Phe - **Glu** - **Glu** - Phe - Val - Leu - Val - Asn - Asp - Gly - Gly - Thr - Gly - **Asp** -
f_2 Ala - Ser - Asn - Phe - **Thr** - **Gln** - Phe - Val - Leu - Val - Asn - Asp - Gly - Gly - Thr - Gly - **Asn** -

20 25 30
fr Val - **Lys** - Val - Ala - Pro - Ser - Asn - Phe - Ala - Asn - Gly - Val - Ala - Glu - Try - Ile - Ser -
f_2 Val - **Thr** - Val - Ala - Pro - Ser - Asn - Phe - Ala - Asn - Gly - Val - Ala - Glu - Try - Ile - Ser -

35 40 45 50
fr Ser - Asn - Ser - Arg - Ser - Gln - Ala - Tyr - Lys - Val - Thr - Cys - Ser - Val - Arg - Gln - Ser -
f_2 Ser - Asn - Ser - Arg - Ser - Gln - Ala - Tyr - Lys - Val - Thr - Cys - Ser - Val - Arg - Gln - Ser -

55 60 65
fr Ser - Ala - **Asn** - Asn - Arg - Lys - Tyr - Thr - **Val** - Lys - Val - Glu - Val - Pro - Lys - Val - Ala -
f_2 Ser - Ala - **Gln** - Asn - Arg - Lys - Tyr - Thr - **Ile** - Lys - Val - Glu - Val - Pro - Lys - Val - Ala -

70 75 80 85
fr Thr - Gln - **Val** - **Gln** - Gly - Gly - Val - Glu - Leu - Pro - Val - Ala - Ala - Try - Arg - Ser - Tyr -
f_2 Thr - Gln - **Thr** - **Val** - Gly - Gly - Val - Glu - Leu - Pro - Val - Ala - Ala - Try - Arg - Ser - Tyr -

90 95 100
fr **Met** - Asn - **Met** - Glu - Leu - Thr - Ile - Pro - **Val** - Phe - Ala - Thr - **Asx** - **Asp** - Asp - Cys - **Ala** -
f_2 **Leu** - Asn - **Leu** - Gln - Leu - Thr - Ile - Pro - **Ile** - Phe - Ala - Thr - **Asn** - **Ser** - Asp - Cys - **Glu** -

105 110 115
fr Leu - Ile - Val - Lys - Ala - Leu - Gln - Gly - **Thr** - **Phe** - Lys - **Thr** - Gly - **Ile** - **Ala** - **Pro** - **Asn** -
f_2 Leu - Ile - Val - Lys - Ala - Met - Gln - Gly - **Leu** - **Leu** - Lys - **Asp** - Gly - **Asn** - **Pro** - **Ile** - **Pro** -

120 125 129
fr **Thr** - Ala - Ile - Ala - Ala - Asn - Ser - Gly - Ile - Tyr
f_2 **Ser** - Ala - Ile - Ala - Ala - Asn - Ser - Gly - Ile - Tyr

FIG. 6-10. **Amino acid sequences of the coat proteins of RNA coliphages fr and f2. Differences are indicated by bold type. (From Hohn and Hohn, 1970.)**

differences in 62 residues can be accounted for by assuming single nucleotide changes in the coat-protein gene of the viral RNAs.

Robinson and associates were able to relate directly nucleotide sequences of phage R17-RNA and amino acids at the N terminus of the R17 coat protein. This was accomplished by digesting the RNAs of R17 and an amber mutant with ribonuclease T1 and comparing the oligonucleotide fractions obtained. Only one was different, and upon sequencing the corresponding oligonucleotide from the wild type phage RNA, it was found to have the proper array of codons to be the first portion of the coat-protein gene.

References

BOOKS

Drake, J. W.: *The Molecular Basis of Mutation,* Holden-Day, San Francisco, 1970.

Hollaender, A. (ed.): *Chemical Mutagens: Principles and Methods for Their Detection,* vol. 1, Plenum, 1971.

Knight, C. A.: *Chemistry of Viruses* (Protoplasmatologia, vol. 4, 2), Springer-Verlag, Vienna, 1963.

Stahl, F. W.: *The Mechanics of Inheritance,* 3d ed., Prentice-Hall, Englewood Cliffs, N.J., 1969.

Stent, G. S.: *Molecular Genetics: An Introductory Narrative,* Freeman, San Francisco, 1971.

The Genetic Code, vol. 31, Cold Spring Harbor Symp. Quant. Biol., Cold Spring Harbor, N.Y., 1966.

JOURNAL ARTICLES AND REVIEW PAPERS

Brenner, S., L. Barnett, F. H. C. Crick, and A. Orgel: The Theory of Mutagenesis, *J. Mol. Biol.* **3**:121–124, 1961.

Funatsu, G., and M. Funatsu: Chemical Studies on Proteins from Two Tobacco Mosaic Virus Strains, *Phytopathol. Soc. Japan,* 1–9, 1968.

Ghendon, Y. Z.: Conditional-lethal Mutants of Animal Viruses, *Prog. Med. Virol.* **14**:68–122, 1972.

Hennig, B., and H. G. Wittmann, Tobacco Mosaic Virus: Mutants and Strains, in *Principles and Techniques in Plant Virology,* C. I. Kado and H. O. Agrawal (eds.), pp. 546–594, Van Nostrand Reinhold, New York, 1972.

Hohn, T., and B. Hohn: Structure and Assembly of Simple RNA Bacteriophages, *Adv. Virus Res.* **16**:43–98, 1970.

Knight, C. A., and W. M. Stanley: Aromatic Amino Acids in Strains of Tobacco Mosaic Virus and in the Related Cucumber Viruses 3 and 4, *J. Biol. Chem.* **141**:39–49, 1941.

Robinson, W. E., R. H. Frist, and P. Kaesberg: Genetic Coding: Oligonucleotide Coding for First Six Amino Acid Residues of the Coat Protein of R17 Bacteriophage, *Science* **166**: 1291–1293, 1969.

Sengbusch, P., von, and H. G. Wittmann: Serological and Physicochemical Properties of the Wild Strains and Two Mutants of Tobacco Mosaic Virus with the Same Amino Acid Exchange in Different Positions of the Protein Chain, *Biochem. Biophys. Res. Commun.* **18**:780–787, 1965.

Siegel, A., and M. Zaitlin: Defective Plant Viruses, in *Perspect. Virol.*, M. Pollard (ed.), vol. IV, pp. 113–124, Harper & Row, New York, 1965.

Streisinger, G., Y. Okada, J. Emrich, J. Newton, A. Tsugita, E. Terzaghi, and M. Inouye: Frameshift Mutations and the Genetic Code, *Cold Spring Harbor Symp. Quant. Biol.* **31**:77–84, 1966.

Terzaghi, E., Y. Okada, G. Streisinger, J. Emrich, M. Inouye, and A. Tsugita: Change of a Sequence of Amino Acids in Phage T4 Lysozyme by Acridine-induced Mutations, *Proc. Natl. Acad. Sci. U.S.A.* **56**:500–507, 1966.

Wittmann-Liebold, B., and H. G. Wittmann: Coat Proteins of Strains of Two RNA Viruses: Comparison of Their Amino Acid Sequences, *Mol. Gen. Genet.* **100**:358–363, 1967.

CHAPTER 7 ORIGIN, RECONSTITUTION, AND SYNTHESIS OF VIRUSES; PROVIRUSES, VIROGENES, AND VIROIDS

The origin of life is unknown. Likewise the origin of viruses, which are considered by some to be a form of life, is obscure. However, an accumulation of knowledge about viruses and cells, especially on the molecular level, has provided a new basis for considering the plausibility of various hypotheses, even though definitive judgment appears remote. Thus, the test-tube synthesis of viral nucleic acids and proteins and the smooth spontaneous assembly in vitro of several viruses lend support to the possible development of viruses by organic chemical evolution. On the other hand, the demonstration that non-nuclear nucleic acids and fragments of nucleic acid are often functional in cells suggests the possibility that viruses might be residual forms of a retrograde evolution, or that they might represent fugitive chromosomes or chromosomal segments. Finally, evidence for unusual entities with viral characteristics, such as viroids, raises new possibilities for the origin and nature of viruses.

7-1 *Origin of Viruses*

IT CAN BE POSTULATED THAT VIRUSES AROSE BY CHEMICAL EVOLUTION, RETROGRADE EVOLUTION, OR ESCAPE OF GENETIC ELEMENTS FROM THEIR USUAL ENVIRONMENT

Students of the origin of life on earth have developed a scheme which goes like this. The earth took form about 5 billion years ago and was surrounded by an atmosphere of gases, chief among which were methane, ammonia, hydrogen, and water vapor. Chemical evolution began as these primitive gases were exposed to thermal energy from the sun and volcanoes, radiation from the sun, and electric discharges (i.e., lightning). Probably ultraviolet light constituted the most abundant source of energy. In any case, it is supposed that the compounds of great importance in modern biologic systems were produced by chemical evolution and were spontaneously organized into a primitive self-reproducing form of life. This primitive form of life, appearing 2 to 3 billion years after the earth took shape, underwent biologic evolution through ages of time to produce the diverse forms of life confronting us at present. In this plan, one would speculate that a virus was the most primitive form of life. Since there were as yet no other forms of life to parasitize, this protovirus would have had to prey on the environment, organizing the available chemicals into replicates of itself. Through an evolutionary process leading to more complex forms, some of the early virus could be imagined to develop into a primitive cell. From this point biologic evolution and natural selection would proceed to give rise to a diversity of organisms.

Reconstruction experiments, starting with the raw materials and kinds of energy postulated to abound around primitive planets, have yielded amino acids, peptides, sugars, lipids, purine and pyrimidine bases, and many other organic compounds (see reviews by Fox et al., 1970, 1971; Ponnamperuma and Gabel, 1968). Even nucleotides have been obtained in low yields and, most amazingly, protein-like compounds (called *polyamino acids* or *proteinoids*) which form specific complexes with various polynucleotides. Fox has demonstrated that his proteinoids are readily converted in hot water to spheroidal structures, about 1 to 4 microns in diameter (Fig. 7-1), which possess a surprising number of features of cells, including catalytic (enzymatic) activities. Such protocells, rather than a primitive virus, may in fact have developed into the first "living" cell.

Another notion of the origin of viruses is that they are descen-

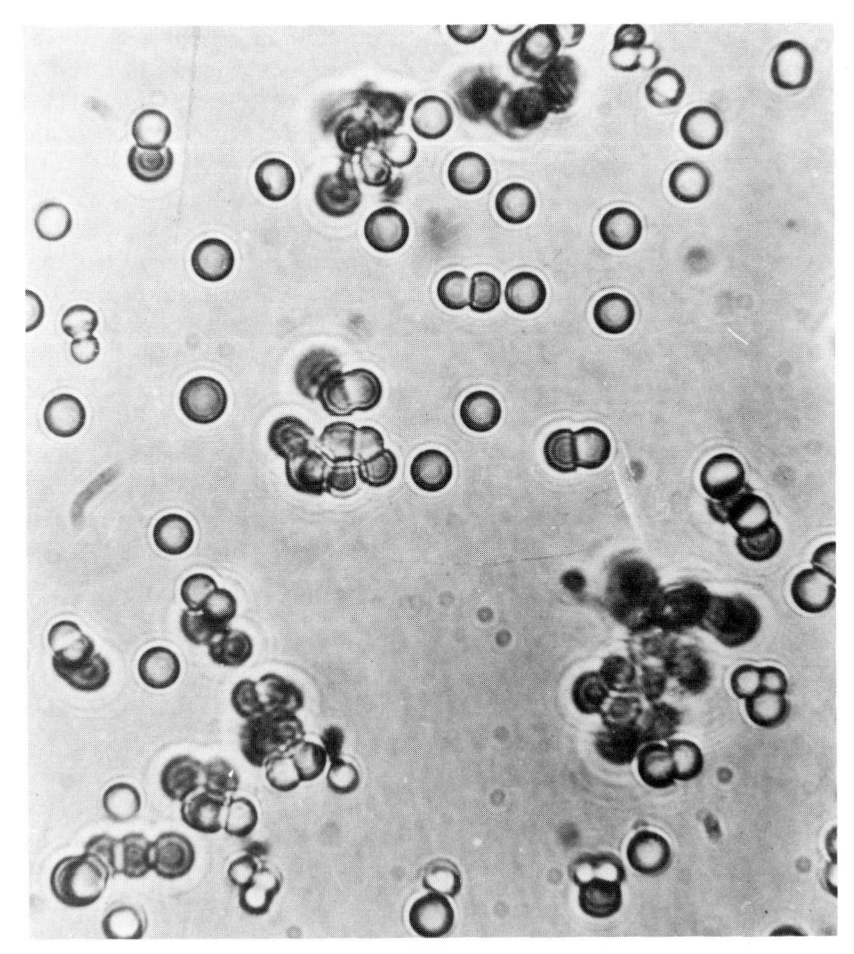

FIG. 7-1. **Protocell models: microspheres produced by heating proteinoid in dilute salt for a short period of time. (Courtesy of S. W. Fox.)**

dants of higher organisms from which they came by retrograde or regressive evolution. For example, there is a progressive decrease in morphologic and metabolic complexity in the series common bacteria, *Bedsonia, Rickettsia,* large viruses, small viruses, and infectious nucleic acid. It appears that starting with autotrophic bacteria, a gradual loss of enzymes and functions could lead ultimately to a completely obligate parasite, one of the currently simplest viruses. However, little evidence has yet been produced to support such a progression; hence it remains a purely speculative idea.

A third notion of the possible origin of viruses is that they were derived from cells not by retrograde evolution but simply as chromosomes or segments of chromosomes that have escaped from their usual habitat. (To account for RNA-containing viruses one would have to assume that a transcribed form of DNA, i.e., RNA, had some selective advantage in certain environments.) Suggestive examples come from bacterial systems. The process of transduction in which phage DNA carries a piece of bacterial chromosome from one cell and establishes it functionally in another is an illustration of the viability of a chromosomal fragment when transferred into another cell. However, a virus is already present here, and the transduced bit of bacterial DNA is scarcely a primitive virus. Much more virus-like qualities are possessed by bacterial plasmids. These are supernumerary DNAs, about one-hundredth the size of the bacterial chromosome (M.W. about 20×10^6), which exhibit a variety of functions and accordingly are called sex factors, resistance transfer factors, and bacteriocinogenic factors. The sex factor can exist in either of two states. In one of these states it is inserted into the bacterial chromosome at one of many possible sites, and upon conjugation of bacterial cells, it promotes the oriented transfer of the bacterial chromosome into the second cell. In the second state, sex factor exists free of the bacterial chromosome in the cytoplasm of the cell, where it replicates autonomously; it promotes conjugation and transfer of material, including itself but not the bacterial chromosome, into the receptor cell, which previously did not contain sex factor. *Escherichia coli* sex factor can be transmitted to and propagated in *Shigella* bacteria as well, but in contrast to the situation in *E. coli,* it does not integrate but occurs as a replicating cytoplasmic entity, capable of promoting conjugation and its own transfer to another cell.

The other plasmids mentioned share with the sex factor the common property of promoting conjugation and genetic transfer between bacteria. In a sense this makes them all sex factors with additional distinctive functions. Thus the resistance transfer factor is noted for its capacity to acquire determinants of resistance to drugs and to transfer them to previously drug-susceptible cells. A determinant in the case of the drug penicillin might be a gene for the penicillin-destroying enzyme, penicillinase. Similarly, bacteriocinogenic bacteria produce bactericidal agents (called bacteriocins, or in the case of those produced by *E. coli,* colicins), and the capacity to produce these lethal antibiotics resides in plasmid DNA, which can be transferred to other cells by conjugation.

Striking similarities are noted when these plasmids are compared with certain bacterial viruses such as the temperate coliphage, lambda. They have DNA genomes of about the same size, and both genomes are independent of the bacterial chromosome in the sense that they can replicate autonomously and if lost from the cell can be reacquired only by infection. Infection by lambda is by means of an injection mechanism, as described in Chap. 4; infection by plasmids is by conjugation which they promote. Both can mutate. Both, under appropriate conditions, can become integrated into the host cell chromosome, in which state they replicate coordinately with the host DNA. Hayes summarizes a discussion of sex factors as follows:

> They are infectious, they depend on the host cell for their metabolism, and they possess only one type of nucleic acid. We think it's logical to regard them as viruses which, instead of killing their host cells and elaborating a complex protein to protect their nucleic acid from the environment and to ensure its infectivity, have developed the conjugation mechanism as an elegant way of promoting their efficient dissemination among a wide range of bacterial species and genera.

However, the key question in considering the relevance of plasmids to the possible origin of viruses is: How did plasmids originate? This is not known, and hence it cannot be claimed that plasmids are fugitive bits of bacterial chromosome with viral properties. Nevertheless, the hypothesis remains attractive because of the homology between plasmid and bacterial chromosome DNAs, which presumably is the basis for the integration of some plasmids. The degree of homology observed would be expected if the plasmids had evolved through many generations from a fugitive segment of bacterial chromosome.

7-2 Reconstitution

NUMEROUS VIRUSES CAN BE WHOLLY OR PARTIALLY RECONSTITUTED AS INFECTIOUS ENTITIES BY BRINGING TOGETHER THEIR CONSTITUENT PROTEINS AND NUCLEIC ACIDS

Takahashi and Ishii found that excess viral coat protein isolated from tobacco leaves infected with tobacco mosaic virus could be polymerized in the test tube into rodlike particles closely resembling the virus, simply by lowering the pH a little. It was only a step from this observation to the copolymerization of viral protein and viral RNA. This was reported a few years later by Fraenkel-Conrat and Williams, who used protein and nucleic acid obtained by disaggregating the virus.

The components for reconstitution of TMV and its strains are usually obtained in two separate reactions, since the best nucleic acid preparations result from a procedure that greatly denatures the protein. Thus, nucleic acid is obtained by extracting the virus with 80 percent phenol (RNA goes into the aqueous phase and protein into the phenolic phase), while the protein is obtained by treating the virus with cold 67 percent acetic acid (the RNA precipitates out while the protein remains in solution). The viral RNA thus obtained is in the form of single strands, and the protein, at around neutral pH, is largely in the form of quasi-stable aggregates comprised of three subunits, generally called *A protein,* together with small amounts of other aggregates, including especially a disk-shaped component consisting of 34 subunits. The latter has been found by Richards and Williams to be essential for rod initiation, while A protein is responsible for rod elongation. The reconstitution reaction with TMV (or its strains) can be represented as follows:

$$\text{A protein} + \text{RNA} \xrightarrow[\text{pH 7.3}]{\text{1 hr at 30°C}} \text{reconstituted virus}$$

A protein	+ RNA		reconstituted virus
M.W. ~ 53,000	M.W. ~ 2×10^6		M.W. 40×10^6
Noninfectious	Slightly infectious		18×300 nm rods
			Highly infectious

As indicated, the protein used in reconstitution has no infectivity, and the slight infectivity of the RNA usually amounts to about 0.05 percent that of the same amount of RNA in a virus particle. In contrast, the reconstituted virus shows infectivities ranging from 30 to 100 percent of those of the virus from which parts were obtained for the reconstitution. In structure as revealed by electron microscopy, in ultraviolet absorbance, and in stability to heat and various pH values, the reconstituted virus is virtually indistinguishable from the virus produced in infected plants (Fig. 7-2). On the other hand, the rodlike particles which can be produced by polymerizing TMV protein alone, though they closely resemble complete virus particles in structure, are devoid of infectivity and are much

FIG. 7-2. **Electron micrographs of a reconstitution mixture of tobacco mosaic virus A protein and tobacco mosaic virus RNA at various times after mixing. (a) Immediately after mixing: only polystyrene reference spheres are visible; (b) 2 min after mixing: many fibrous nucleic acid par-**

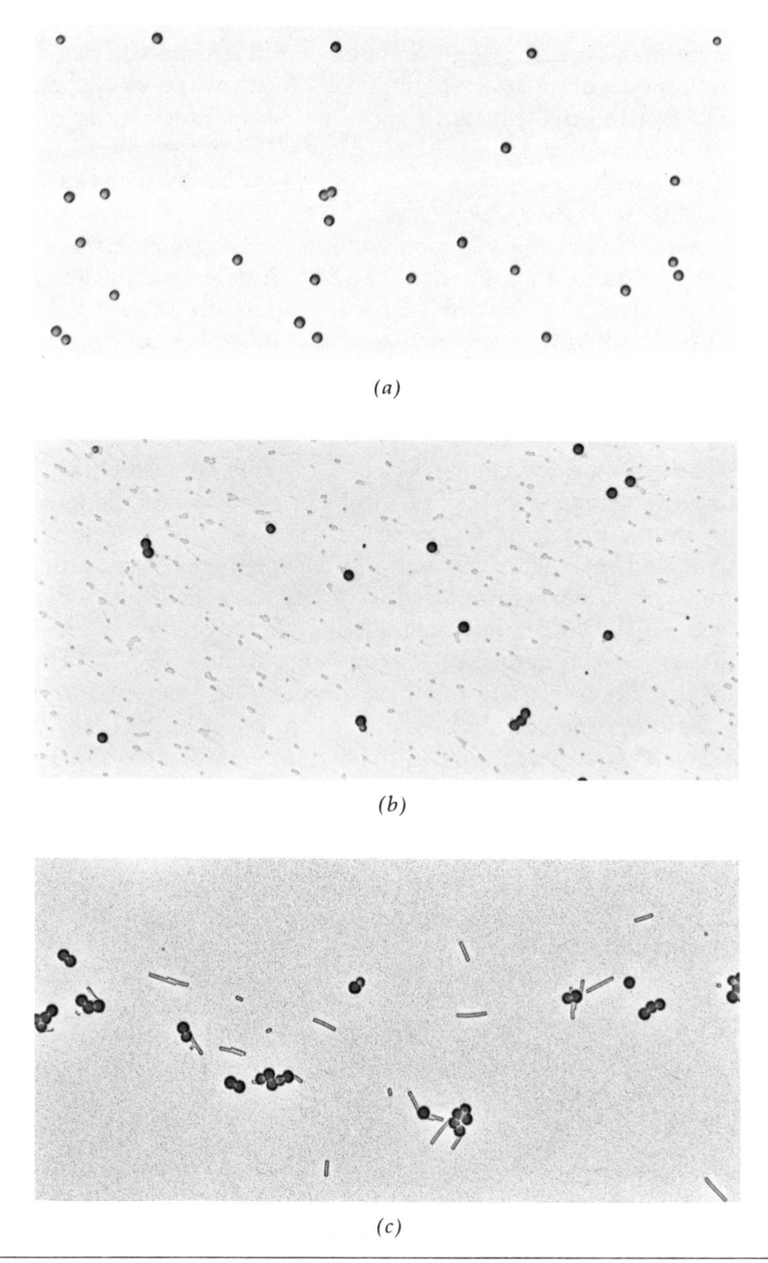

ticles are visible, each of which has some protein assembled around one end representing an early stage in the reconstitution of viral rods; (c) 6 hr after mixing: many full-sized TMV particles are present. (Courtesy of K. Richards.)

less stable than virus. Taking the rodlike particles formed from TMV protein alone to pH 8 to 9 disaggregates them, while virus particles isolated from infected plants or those of reconstituted virus remain intact. It appears that the rods containing RNA are stabilized by the RNA (presumably by formation of secondary bonds between RNA and protein).

A significant aspect of reconstitution in the TMV series is that mixed reconstitutions are possible between proteins and nucleic acids of different strains. An example of this is given in Table 7.1. It can be seen by comparing the homologous and heterologous viruses that the symptom expression in plants depends solely on the nucleic acid. Furthermore, infection of plants with heterologous reconstituted viruses yields homologous progeny, with the protein matching the nucleic acid of the infecting species. This kind of result helped to establish the genetic function of viral nucleic acid and its independence from the protein in this regard.

Though the type of response in infected plants in terms of symptoms and products produced is clearly dependent upon the nucleic acid present in the infecting species, the infectivity is considerably altered by the nature of the protein coat (Table 7.2). Thus it appears that the infectivities of heterologously reconstituted virus in which the coat protein is from common TMV can be 10 times those of the homologous intact viruses (last four examples in Table 7.2). This surprising result is probably best explained by the assumption

TABLE 7-1 **Homologous and Heterologous Reconstitution with the Nucleic Acids and Proteins of Common Tobacco Mosaic Virus (TMV) and Yellow Aucuba Strain (YA)**

			Symptoms produced by infection of:		*Viral components in progeny*	
Protein	*RNA*	*Reconstituted virus*	*Turkish tobacco*	Nicotiana sylvestris	*Protein*	*RNA*
TMV	TMV	TMV	Green mosaic	Green mosaic	TMV	TMV
YA	YA	YA	Yellow mosaic	Brown lesions	YA	YA
YA	TMV	YA protein: TMV RNA	Green mosaic	Green mosaic	TMV	TMV
TMV	YA	TMV protein: YA RNA	Yellow mosaic	Brown lesions	YA	YA

TABLE 7-2 **Infectivity of Homologous and Heterologous Reconstituted Viruses in the Tobacco Mosaic Virus Series**

Source of RNA reconstituted with protein of common TMV	*Number of lesions* produced on xanthi tobacco by equivalent amounts of RNA in the form of:*		
	Isolated RNA	*Intact virus*	*Reconstituted virus*
Common TMV	91 ± 8	9,700 ± 724	10,400 ± 999
Strain 256	75 ± 7	890 ± 118	10,900 ± 893
Strain 257	93 ± 14	860 ± 51	9,900 ± 880
Strain 314	66 ± 8	730 ± 42	10,700 ± 211
Strain HR	74 ± 8	440 ± 58	4,100 ± 543

* From 4 to 10 assays were made for each species tested; the lesion counts are presented in terms of lesions per half-leaf per microgram of substance tested, together with standard deviations from the means.

SOURCE: Adapted from Holoubek, 1962.

that in infected plants TMV protein is more efficiently removed from strains RNAs than the homologous proteins, or conversely, TMV protein is more slowly removed and hence more viral RNA is preserved until it reaches the active site in the cell. In any case, the enhanced infectivities observed suggest that in mixed infections in nature, substantial increases in virulence could occur without mutation simply as a consequence of heterologous combinations of protein and nucleic acids (i.e., phenotypic mixing).

Several natural and synthetic polynucleotides have been reconstituted with TMV protein but in no case with the degree of success achieved in heterologous reconstitutions with strains of TMV. Physically successful reconstitution results in every case in the production of rodlike particles, since, as noted in Chap. 3, it is the protein that determines the structure of a virus particle. The rods of variable length formed with the RNA of turnip yellow mosaic virus (TYMV, a spheroidal virus) and TMV protein possess slight infectivity (in Chinese cabbage, the usual host for TYMV); however, this is lost after treatment with ribonuclease. Since mixed reconstitutions among TMV strains yield ribonuclease-resistant products, it appears that the rods formed with TMV protein and TYMV RNA have a looser structure than those formed from components of TMV strains. Rodlike particles can be obtained with coliphage MS2 RNA and TMV protein and with polyadenylic acid and TMV protein, but neither of these shows any infectivity.

The ability of TMV protein to participate in reconstitution to form stable rodlike particles is apparently dependent on several factors, including the composition of the protein, its size, and certain strategically placed charged groups. As noted in Chap. 6, coat-protein mutants of TMV in which only two amino acids of the 158 were exchanged can be defective in ability to form characteristic particles. The importance of size has been demonstrated by removing amino acid residues from the C-terminal end of a mutant of TMV in which the sequence of amino acids is the same as in TMV except that a leucine has replaced proline in position 156 (see von Sengbusch and Wittmann, 1965). Removal of three C-terminal amino acid residues does not affect the ability of the protein to form rodlike particles, but removal of 15 to 17 amino acids abolishes this capacity.

The influence of charge on the reconstitution reaction can be illustrated as follows. The protein of the common strain of TMV has two lysine residues per subunit (positions 53 and 68), and the epsilon amino groups of these residues are normally positively charged. These charges can be selectively abolished, for example, by trifluoroacetylation. When this is done, the protein is rendered incapable of self-assembly to rodlike structures; removal of the trifluoroacetyl groups restores the ability of the protein to form typical rodlike structures. This suggests that the positive charge on one or both of the two lysine residues is essential for the fabrication of the coat-protein superstructure.

For years tobacco mosaic virus was the only virus reconstituted in the laboratory. Reconstitution of a virus has the prerequisite that disassembly be possible in such a manner as to avoid breaking or irreversibly denaturing the viral constituents. Many spherical viruses appear to be more difficult to disassemble than TMV and are even harder to put back together. However, Bancroft found a series of serologically unrelated, spheroidal (about 26 nm in diameter) plant viruses which could be successfully dissociated if dialyzed overnight in 1 *M* NaCl at pH 7 or a little higher. These viruses were cowpea chlorotic mottle virus (CCMV), brome mosaic virus (BMV), and broad-bean mottle virus (BBMV). Quantitative reconstitution of each of these viruses can be obtained if a stoichiometric mixture of protein (isolated by the salt dissociation method) is mixed with viral RNA (obtained by treating the virus with 80 percent aqueous phenol) and the mixture dialyzed at 4°C against dilute, buffered KCl containing magnesium ions (0.01 *M* KCl, 0.01 *M* tris, pH 7.4, 5×10^{-3} *M* $MgCl_2$). The reconstituted product is as infectious as the virus from which parts were obtained and exhibits the same size and shape, as judged by electron microscopy and sedimentation studies. Furthermore, heterologous reconstitution can be done among pairs of the three viruses to form "hybrids" containing various combinations of RNA and protein which are similar to the products of the natural phenomenon (i.e., phenotypic mixing). All these are infectious, and differences in host specificity and symptoms shown by the original viruses are maintained by the hybrids; in each case, the specificity and symptoms are those of the virus that provides the RNA. This is the same result as observed with mixed reconstitution in the TMV series. These hybrids are also similar to those in the TMV series in that protein of these spheroidal viruses can be used to coat a variety of other RNAs, yielding in each case spheroidal particles.

Thus, the general principles observed in reconstitution of strains of TMV were confirmed by results of the studies made with three spheroidal plant viruses: (1) in reconstituted plant viruses the shape of the particles is determined by the protein component; (2) host specificity and other inheritable characteristics, such as symptoms, are determined solely by the nucleic acid.

Reconstitution has also been achieved with the simple, RNA-containing bacteriophages of the f2 group (e.g., f2, fr, MS2, R17) and with the similar Qβ phage (see Hohn and Hohn, 1970). These phages are spheroidal, with a diameter of about 27 nm, and consist of a molecule of single-stranded RNA in a protein shell of 180 subunits (M.W. 13,750). In addition, one molecule of protein (M.W. 35,000) called *A protein* or *maturation protein* is present. The general problem of reconstitution, namely, getting the viral components in a soluble condition, is severe with these RNA phages. In fact, successful reconstitution requires that the mixing of components be done in denaturing media such as ice-cold dilute acetic acid or in guanidine hydrochloride. Such mixtures, when dialyzed against renaturing buffers, yield particles resembling the in vivo product except that they usually contain a higher proportion of misshaped particles. Essentially no infectivity is observed with the particles obtained from mixtures of viral RNA and the major coat protein, but addition of A protein to the reconstitution mixture results in particles with definite but low infectivity. The A protein appears to be especially insoluble and difficult to manage in these experiments.

Partial reconstitution of structurally more complex coliphages such as T4 and lambda, and *Salmonella* phage P22, has been achieved by mixing phage parts under appropriate conditions (see Wood and Edgar, 1967; Edgar and Lielausis, 1968). Appropriate conditions might be, for example, incubation for a few hours at room temperature in 0.85 percent saline solution buffered with M/150 phosphate at pH 7. Advantage is taken here of conditional lethal mutants which under restrictive conditions (such as elevated temperature) are able to produce some but not all of the phage parts required for infectious particles. Selection, then, of a pair of mutants, which together provide the components of a complete particle, will lead to infectious phage provided that the components will unite spontaneously. For example, a certain mutant of T4 when grown at an elevated temperature produces phage particles which are complete except for tail fibers. Since tail fibers are the attachment organs for T4, the particles lacking these are not infectious. Such defective particles can be isolated from artificial lysates (infected cells caused to burst by treatment with chloroform). Similarly, another mutant defective with respect to phage head formation produces many tail fibers which can be isolated from artificial lysates of the infected cells. These fibers, incubated with the tail fiberless particles, spontaneously combine to form complete, infectious particles.

Poliovirus is a small, spheroidal animal virus composed of one molecule of single-stranded RNA and a protein coat which has four species of proteins. Purified poliovirus can be dissociated into its protein and nucleic acid components by holding it at 25°C for 60 min in 10 *M* urea and 0.1 *M* mercaptoethanol (the latter preserves —SH groups in proteins which might otherwise interact to form —S—S— cross-linkages between polypeptide chains, often rendering them insoluble). Infectious poliovirus can be reconstituted from its parts by a five-step dilution of the dissociation mixture in cold phosphate-buffered saline solution at pH 7.2. Attempts to reconstitute by a one-step dialysis of the dissociated virus have been unsuccessful.

7-3 Laboratory Synthesis of Viruses

ENZYMATIC SYNTHESES OF INFECTIOUS VIRAL RNA AND INFECTIOUS VIRAL DNA HAVE BEEN ACCOMPLISHED IN THE LABORATORY

The next step beyond reconstituting viruses from their nucleic acid and protein constituents is the synthesis of infectious viral nucleic acids in cell-free media starting from low molecular weight chemicals such as nucleotides. This was first reported in 1965 for the RNA of coliphage Qβ (see Spiegelman et al., 1967).

When Qβ is grown in *E. coli,* strain Q13, a virus-coded RNA polymerase, also called *replicase,* is produced in significant quantities along with phage. This replicase can be isolated and purified by column chromatography. Synthesis of RNA by replicase can be tested for by adding the purified enzyme to a mixture of the four nucleotide triphosphate precursors of RNA, Mg^{2+}, pH 7.4 buffer, and some nucleic acid as a template. The amount of synthesis can be estimated in terms of counts incorporated in acid-insoluble polynucleotide if one of the nucleotide triphosphates is radioactively labeled (e.g., ^{32}P-labeled UTP). The Qβ replicase was found to have a high specificity for Qβ-RNA template, as shown by results such as those summarized in Table 7.3. In some experiments as much as a 100-fold increase of product over Qβ template RNA was observed. The product is noninfectious for Q13 *E. coli* cells, in accordance with the behavior of template RNA. On the other hand, the product is infectious in Q13 spheroplasts (spheroplasts are bacteria stripped of cell walls by lysozyme) and appears to contain the complete phage genome, since the progeny from spheroplast infection are able to infect whole Q13 cells. Furthermore, the product of replicase action in response to Qβ-RNA template is itself a good template, as was shown by its ability to carry the synthesis serially through 15 tubes without requiring any fresh exogenous template. Product produced in the fifteenth tube was as infectious as in the initial tube when tested in spheroplasts.

Specificity in this synthesis resides predominantly in the tem-

TABLE 7-3 **Incorporation of Nucleotide Radioactivity into Polynucleotide by Qβ-Replicase Action in Response to Various Templates**

Templates	*Counts/min incorporated**
Qβ phage RNA	4,929
Turnip yellow mosaic virus RNA	146
Tobacco necrosis satellite RNA	61
Ribosomal RNA	45
MS2 phage RNA	35

* Control reactions containing no template yielded an average of 30 counts per minute.

SOURCE: From Spiegelman et al., 1967.

plate rather than in the replicase. This was shown by using replicase from one strain of Qβ and RNA from a mutant. The product RNA was found to have the same properties as the template RNA, rather than those of the RNA of the phage which coded for the replicase.

It will be noted that this laboratory synthesis of infectious viral RNA was not complete, in the sense that some viral RNA was essential as a template. A complete synthesis would require putting a sequence of nucleotides together without benefit of template to produce an infectious polynucleotide. This may eventually be done.

Similarly, an enzymatic synthesis of infectious coliphage ϕX174 DNA can be accomplished (see Goulian et al., 1967), although it is considerably more complicated than the Qβ-RNA situation. The genome of phage ϕX174 is a small, circular, single-stranded DNA molecule. When such nucleic acid (which can be designated "+ strand") is isolated and treated with a purified DNA polymerase from *E. coli* (often called the "Kornberg enzyme") in the presence of appropriate nucleotide triphosphates, a complementary (minus) strand is produced but not closed into a ring structure. Treatment of this product with a polynucleotide ligase enzyme closes the ring, to yield a double-stranded DNA ring. (This has also been identified as a form that occurs during infection of cells by ϕX174, where it has been called "replicative form" or "RF"). If the double-stranded DNA ring is now subjected to limited action of pancreatic deoxyribonuclease, a variety of structures is produced, among which are some rings in which the plus strand has been opened while the minus strand is still in a closed-ring form. Heating under proper conditions separates these two species, and the intact rings of minus strand can be isolated by density-gradient centrifugation. The minus strands are then used as templates for the DNA polymerase, which produces complementary plus strands that are subsequently closed into rings by the ligase enzyme. The final product is a double-stranded DNA ring structure in which both strands have arisen from cell-free, laboratory synthesis. Both synthetic minus circles and synthetic RF were found to be infectious for bacterial spheroplasts. Plus circles can be isolated from synthetic RF by treatments similar to those outlined for obtaining minus circles in one of the earlier steps. These, like the plus circles extractable from virus particles, are also infectious when tested in spheroplasts. Figure 7-3 summarizes major steps and products of the synthesis.

Though the cell-free production of infectious ϕX174 DNA can be considered another milestone among biochemical syntheses, it, like the Qβ-RNA instance, does not represent a complete synthesis, since it cannot be done without some viral DNA added as template.

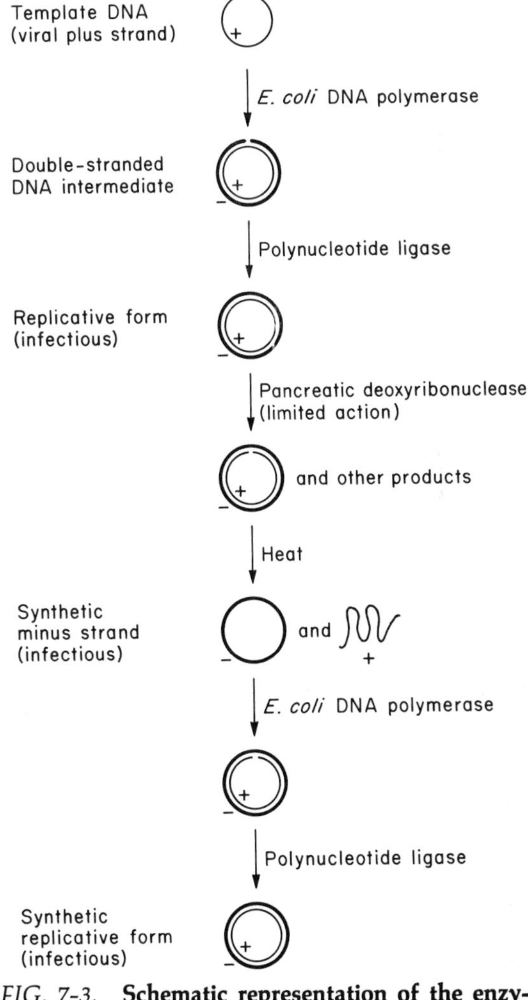

FIG. 7-3. **Schematic representation of the enzymatic synthesis of infectious phage φX174 DNA. (Adapted from Goulian et al., 1967, but simplified by omitting important but distracting details such as radioactive and density labeling and some of the products.)**

7-4 *Proviruses and Virogenes*

RNA TUMOR VIRUSES ARE DISTINCTIVE AMONG RNA-CONTAINING VIRUSES IN THAT THEY APPEAR TO FUNCTION THROUGH AND TO PERPETUATE THEMSELVES WITH SPECIFIC DNA. IN ONE HYPOTHESIS IT IS PROPOSED THAT THIS DNA IS A PROVIRUS FORM INDUCED BY INFECTION WITH EXOGENOUS RNA, WHILE IN ANOTHER IT IS POSTULATED THAT GENES FOR RNA TUMOR VIRUSES (VIROGENES) HAVE BEEN PRESENT FOR AGES IN THE NUCLEAR DNA OF ALL VERTEBRATES

RNA tumor viruses (also called *leukoviruses*) have been known for years as causes of malignant growths of blood and lymph cells (leukemias, erythroblastoses, myeloblastoses, lymphomatoses, etc.) and occasionally of hard fleshy tumors (sarcomas, carcinomas, etc.) in chickens and mice. RNA tumor virus particles are roughly spheroidal and are about 100 nm in diameter. The particles have a central core, or nucleoid, containing 10 to 12 million daltons of single-stranded RNA and some protein, including a group-specific antigen; these constituents are surrounded by an envelope containing some host material and viral type-specific antigens. Avian and murine leukovirus particles of the sort just described are sometimes referred to as "C-type particles," according to electron-microscope distinctions made by Bernhard. In this classification, B-type particles are about the same size as C particles but the cores are eccentrically located (mouse mammary tumor virus particles are of this type), while A-type particles are like C-type only smaller, about 70 nm in diameter. C-type RNA viruses have been isolated from or observed in tumors of cats, rats, hamsters, snakes, cows, pigs, monkeys, man, and, most frequently, chickens and mice.

The relationship between leukoviruses and their hosts is complex. Sometimes tumors are present without detectable infectious virus; in other cases, leukemia or tumors may be evident, as well as much infectious virus. In addition, a third case is common in which much virus is demonstrable with little or no evidence of disease. For instance, whole flocks of chickens can be infected with avian leukosis virus, and yet most of the birds remain symptomless all their lives.

The transmission of leukoviruses is also bizarre. Instead of the expected spread of infectious virus mainly from individual to individual (horizontal transmission), it seems more common for RNA tumor viruses to be passed congenitally from mother to progeny (vertical transmission). In the case of cell cultures, vertical transmission is often manifested by transfer of virus (or the potential to produce virus) from cell to cell by cell division, rather than by release from infected cells and reinfection of new cells.

Clues to the possible mechanism of leukovirus operation are contained in the following observations. Replication of RNA tumor viruses requires early DNA synthesis and is sensitive to actinomycin D (this drug usually inhibits host cell DNA-dependent RNA synthesis but not RNA-dependent RNA synthesis). Most leukoviruses appear to have an RNA-dependent DNA polymerase (reverse transcriptase) and perhaps other enzymes concerned with DNA synthesis.

Upon consideration of these and other facts, Temin evolved a provirus hypothesis, which can be sketched as follows. RNA tumor virus infects a cell in the usual way, involving attachment, penetration by engulfment, and dissociation of the particle to release viral RNA. Information is then transferred from RNA to DNA, the provirus, through the mediation of reverse transcriptase. The DNA provirus is presumed to act as a template for the production of the viral RNA component when conditions favor multiplication of the virus, but if virus is not being replicated the DNA replicates once when an infected cell divides, thus disseminating virus information among the daughter cells. It can be imagined that all or a part of the provirus may be integrated into host DNA, where it would be somewhat analogous to the prophage of lysogenized bacterial cells. In some instances perhaps only the viral genes capable of causing neoplastic transformation of cells get incorporated into host genome. However, evidence for the existence of DNA provirus is still indirect, and there are many loose ends in the provirus hypothesis.

In the virogene hypothesis proposed by Todaro and Huebner, it is suggested that the cells of most or all vertebrate species contain genomes of C-type RNA tumor viruses integrated in the form of complementary DNA into host DNA. In this form, viral information is transferred from parent to offspring for countless generations. It is postulated that the integrated viral genes need not be located together in the host cell DNA but may even be dispersed among different chromosomes. Among the virogenes are those concerned with the capacity to cause malignant transformation of cells, which consequently have been termed *oncogenes*. Control of expression of virogenes is presumed to reside in a repressor system, and it can be supposed that selective derepression of virogenes would permit a variety of viral functions to be expressed. For example, virogenes responsible for a tumor antigen might be the only ones operative, or the oncogenes alone might be functional. The extreme cases would be ones in which no virogenes would be functional and in which all of them are functional, respectively. Thus the state of function of virogenes becomes crucial, and it is argued that cancer biologists should pay less attention to the process of infection with tumor viruses and concentrate more on mechanisms of gene regulation, with special emphasis on how oncogenes are turned on and off. As a unifying hypothesis for cancer, it is suggested that various environmental factors may cause derepression of latent oncogenes. Mutation, radiation, chemical carcinogens, cellular constituents, and even oncogenic DNA viruses may serve as such factors.

Some of the principal lines of support cited for the virogene hypothesis are: (1) the widespread occurrence of C-type viruses in vertebrates as determined by isolation of such viruses, by observation with the electron microscope of C-type particles in tumor tissues or cells, or by the detection of characteristic viral antigens; (2) the spontaneous appearance of C-type RNA viruses in certain presumably uninfected mouse embryo cells after several months of culture and many transfers; (3) the induction of infectious virus in randomly selected populations of rodent (mouse, rat, hamster) embryo cells by such agents as x-ray or ultraviolet light. Virus appears even though previously the same cells gave no indication of the presence of virus by infectivity assays, tests for viral antigens, and search for C-type particles.

The points just enumerated could also be used to support the provirus hypothesis, for the main distinction between virogene and provirus appears to be the postulated time at which cells acquired these genetic elements by infection. This is recent for provirus and ancient for virogene, although it is also suggested that virogenes may be endogenous parts of vertebrate cells, rather than a consequence of infection. In either case, more evidence, direct or indirect, is being sought for the existence of provirus-virogene DNA and for its character and mode of function.

7-5 *Viroids*

SINGLE-STRANDED RIBONUCLEIC ACID WITH MOLECULAR WEIGHT AROUND 10^5 APPEARS TO CAUSE DISEASE IN SOME PLANTS

A disease causing spindle-shaped potatoes and hence called "spindle tuber condition" was reported in 1922 by W. H. Martin in a bulletin of the New Jersey State Potato Association. For years thereafter the disease was considered a plant virus disease that was perpetuated from infected tubers during the vegetative propagation of potatoes, and was transmitted mechanically from infected to healthy leaves during growth and cultivation of the plants, and by insect vectors such as aphids. The disease could also be transmitted by grafting. However, early attempts to work with and to isolate the virus were fraught with difficulties, some of which were reduced when it was found that the virus multiplies in tomato plants, where the symptoms are more readily detected than in potato plants. Tomato plants infected with the spindle tuber virus become noticeably stunted and the leaves develop a downward curvature (epinasty), a slight wrinkling (rugosity), and mottling, and often show some purplish-brown necrosis of midribs and lateral veins. These symptoms are not uncommon in plant virus diseases, and the means of transmission of the spindle tuber agent are basically similar to those of many plant viruses. However, the presence of typical virus particles in buffer extracts of infected plants is not demonstrable by sedimentation or electron microscopy, and attempts to demonstrate that infected cells synthesize any viral coat protein have been unsuccessful. On the other hand, buffer extracts of infected plants contain very slowly sedimenting, ribonuclease-sensitive, infectious material. Diener and his colleagues concluded that this infectious agent is small, free RNA. A similar infectious RNA appears to be the form in which the virus of exocortis disease of citrus occurs; in fact, recent evidence suggests that the exocortis and spindle tuber agents may be either identical or strains of the same agent.

Diener proposed the term *viroid* for such agents, thus indicating their virus-like properties but suggesting basic differences. The distinctive features of viroids are their nucleic acids, which are much smaller than those of viruses, and their lack of protein coats. It has not yet been possible to characterize viroids in chemical detail because they occur in such small amounts in infected plants that even the most highly purified preparations consist mostly of host RNA. The RNA composition of viroids is inferred from the abolishment of infectivity upon treatment of purified preparations with ribonuclease but not with deoxyribonuclease. The fact that treatment of the preparations with phenol and other organic chemicals does not alter infectivity argues against involvement of proteins in viroid structure. The size of viroid RNA is determinable only by correlating infectivity with physically separated fractions of nucleic acid whose approximate sizes can be estimated by comparison with standard RNA markers. Thus, fractions can be collected from different depths in the tube after purified viroid has been centrifuged in a sucrose density gradient. The viroid is located by testing the fractions for infectivity, and size is then estimated by relating the distance that infectivity is sedimented to the distance that marker RNAs of known size sediment under the same conditions. Similarly, the viroid preparation can be subjected to electrophoresis in acrylamide gel, and the position to which infectivity migrates is related to the distance that reference RNAs of known size migrate under the same conditions. It can be concluded that viroids are single-stranded RNA because infectivity elutes from hydroxyapatite columns at phosphate concentrations and temperatures known to elute single-stranded but not double-stranded RNA. Furthermore, the eluted infectious species does not induce interference with multiplication of vesicular stomatitis virus in cell cultures, whereas double-stranded RNA from a wide variety of sources does.

From these various tests and measurements, it can be concluded that spindle tuber viroid is single-stranded RNA with a molecular weight of 100,000 or less. Such a small nucleic acid is unprecedented among viruses and raises fundamental questions. For example, how does such a nucleic acid replicate? Does it code for any proteins? If all the nucleic acid served as messenger, the protein coded for would only have a molecular weight of about 12,000. Is this big enough for an RNA replicase enzyme? No replicase this small has yet been identified. How does viroid RNA evoke the various symptoms that disease-infected plants exhibit? There seems not to be enough RNA to provide genes concerned with disease symptoms. The answers to these and other questions await further characterization of viroids. In the meantime, virologists and pathologists are wondering how common such agents are in nature. The possibility has already been suggested that a small DNA molecule may be involved in the disease of sheep called *scrapie,* which has hitherto been thought to be a virus disease. On the other hand, some properties of the scrapie agent suggest that it contains neither protein nor nucleic acid. Resolution of these uncertainties awaits the intensive application of molecular virologic techniques.

References

BOOKS

Campbell, A. M.: *Episomes,* Harper & Row, New York, 1969.

Hayes, W.: *The Genetics of Bacteria and Their Viruses,* 2d ed., Wiley, New York, 1968.

Kimball, A. P., and J. Oro (eds.): *Prebiotic and Biochemical Evolution,* American Elsevier, New York, 1971.

Ponnamperuma, C. (ed.): *Exobiology* (North-Holland Research Monographs, *Frontiers of Biology,* vol. 23), Lange & Springer, Berlin, 1972.
Sager, R.: *Cytoplasmic Genes and Organelles,* Academic, New York, 1972.

JOURNAL ARTICLES AND REVIEW PAPERS

Enzymatic Synthesis of Viral Nucleic Acids

Goulian, M., A. Kornberg, and R. L. Sinsheimer: Enzymatic Synthesis of Infectious Phage ϕX174 DNA, *Proc. Natl. Acad. Sci. U.S.A.* **58**:2321–2328, 1967.
Spiegelman, S., I. Haruna, N. R. Pace, D. R. Mills, D. H. L. Bishop, J. R. Claybrook, and R. Peterson: Studies in the Replication of Viral RNA, *J. Cell. Physiol.* **70**:35–64, 1967.

Origin of Life

Fox, S. W.: Chemical Origin of Cells—part 2, *Chem. Eng. News* **49**:46–53, 1971.
———, K. Harada, G. Krampitz, and G. Mueller: Chemical Origin of Cells, *Chem. Eng. News* **48**:80–94, 1970.
Ponnamperuma, C., and N. W. Gabel: Current Status of Chemical Studies on the Origin of Life, *Space Life Sci.* **1**:64–96, 1968.
Smith, A. E., and D. H. Kenyon: *Is Life Originating de novo? Perspect. Biol. Med.* **15**:529–542, 1972.
——— and ———: The Origin of Viruses from Cellular Genetic Material, *Enzymologia* **43**:13–18, 1972.

Plasmids

See Books: Campbell, 1969; Hayes, 1968; Sager, 1972.
Clowes, R. C.: Molecular Structure of Bacterial Plasmids, *Bacteriol. Rev.* **36**:361–405, 1972.

Reconstitution of Viruses

Bancroft, J. B.: The Self-assembly of Spherical Plant Viruses, *Adv. Virus Res.* **16**:99–134, 1970.
Drzeniek, R., and P. Bilello: Reconstitution of Poliovirus, *Biochem. Biophys. Res. Commun.* **46**:719–724, 1972.
Edgar, R. S., and I. Lielausis: Some Steps in the Assembly of Bacteriophage T4, *J. Mol. Biol.* **32**:263–276, 1968.
Fraenkel-Conrat, H., and B. Singer: Virus Reconstitution: II. Combination of Protein and Nucleic Acid from Different Strains, *Biochim. Biophys. Acta* **24**:540–548, 1957.
——— and R. C. Williams: Reconstitution of Active Tobacco Mosaic Virus from its Inactive Protein and Nucleic Acid Components, *Proc. Natl. Acad. Sci. U.S.A.* **41**:690–698, 1955.
Hohn, T., and B. Hohn: Structure and Assembly of Simple RNA Bacteriophages, *Adv. Virus Res.* **16**:43–98, 1970.

Holoubek, V.: Mixed Reconstitution between Protein from Common Tobacco Mosaic Virus and Ribonucleic Acid from Other Strains, *Virology* **18**:401–404, 1962.

Kushner, D. J.: Self-assembly of Biological Structures, *Bacteriol. Rev.* **33**: 302–345, 1969.

Leberman, R.: The Disaggregation and Assembly of Simple Viruses, *Symp. Soc. Gen. Microbiol.* **28**:183–205, 1968.

Richards, K. E., and R. C. Williams: Assembly of Tobacco Mosaic Virus in vitro: Effect of State of Polymerization of the Protein Component, *Proc. Natl. Acad. Sci. U.S.A.* **69**:1121–1124, 1972.

Takahashi, W. N., and M. Ishii: The Formation of Rod-shaped Particles Resembling Tobacco Mosaic Virus by Polymerization of a Protein from Mosaic-diseased Tobacco Leaves, *Phytopathology* **42**:690–691, 1952.

von Sengbusch, P., and H. G. Wittmann: Serological and Physico-chemical Properties of the Wild Strain and Two Mutants of Tobacco Mosaic Virus with the Same Amino Acid Exchange in Different Positions of the Protein Chain, *Biochem. Biophys. Res. Commun.* **18**:780–787, 1965.

Wood, W. B., and R. S. Edgar: Building a Bacterial Virus, *Sci. Am.* **217**: 60–74, 1967.

RNA Tumor Viruses

Bernhard, W.: The Detection and Study of Tumor Viruses with the Electron Microscope, *Cancer Res.* **20**:712–726, 1960.

Temin, H. M.: The RNA Tumor Viruses—Background and Foreground, *Proc. Natl. Acad. Sci. U.S.A.* **69**:1016–1020, 1972.

Todaro, G. J., and R. J. Huebner: The Viral Oncogene Hypothesis: New Evidence, *Proc. Natl. Acad. Sci. U.S.A.* **69**:1009–1015, 1972.

Viroids

Adams, D. H.: The Scrapie Agent: A Small Deoxyribonucleic Acid-mediated Virus? *Biochem. J.* **127**:82P–83P, 1972.

Diener, T. O.: A Plant Virus with Properties of Free Ribonucleic Acid: Potato Spindle Tuber Virus, in *Comparative Virology*, K. Maramorosch and E. Kurstak (eds.), pp. 433–478, Academic, New York, 1971.

———: Potato Spindle Tuber "Virus." IV. A Replicating Low Molecular Weight RNA, *Virology* **45**:411–428, 1971.

——— and R. H. Lawson: Chrysanthemum Stunt: A Viroid Disease, *Virology* **51**:94–101, 1973.

Lewandowski, L. J., P. C. Kimball, and C. A. Knight: Separation of the Infectious Ribonucleic Acid of Potato Spindle Tuber Virus from Double-Stranded Ribonucleic Acid of Plant Tissue Extracts, *J. Virol.* **8**:809–812, 1971.

Semancik, J. S., and L. G. Weathers: Pathogenic 10S RNA from Exocortis Disease Recovered from Tomato Bunchy-top Plants Similar to Potato Spindle Tuber Virus Infection, *Virology* **49**:622–625, 1972.

INDEX